UNITEXT for Physics

Series editors

Paolo Biscari, Milano, Italy
Michele Cini, Roma, Italy
Attilio Ferrari, Torino, Italy
Stefano Forte, Milano, Italy
Morten Hjorth-Jensen, Oslo, Norway
Nicola Manini, Milano, Italy
Guido Montagna, Pavia, Italy
Oreste Nicrosini, Pavia, Italy
Luca Peliti, Napoli, Italy
Alberto Rotondi, Pavia, Italy

UNITEXT for Physics series, formerly UNITEXT Collana di Fisica e Astronomia, publishes textbooks and monographs in Physics and Astronomy, mainly in English language, characterized of a didactic style and comprehensiveness. The books published in UNITEXT for Physics series are addressed to graduate and advanced graduate students, but also to scientists and researchers as important resources for their education, knowledge and teaching.

More information about this series at http://www.springer.com/series/13351

Franco Strocchi

A Primer of Analytical Mechanics

 Springer

Franco Strocchi
INFN
Theoretical Physics Group at Pisa University
Pisa, Italy

ISSN 2198-7882   ISSN 2198-7890   (electronic)
UNITEXT for Physics
ISBN 978-3-030-08853-8   ISBN 978-3-319-73761-4   (eBook)
https://doi.org/10.1007/978-3-319-73761-4

Printed on acid-free paper

This Springer imprint is published by Springer Nature
The registered company is Springer International Publishing AG
The registered company address is: Gewerbestrasse 11, 6330 Cham, Switzerland

Contents

Preface vii

1   **Difficulties of Cartesian Newtonian Mechanics** 1
    1.1   Constraint forces 1
    1.2   Non-inertial frames and fictitious forces 3

2   **Lagrange equations** 5
    2.1   Degrees of freedom and Lagrangian coordinates 5
    2.2   Lagrangian form of Newton's equations 7
    2.3   Lagrange equations 8
    2.4   Lagrange equations at work. Examples 11
    2.5   Generalized potential 16
    2.6   Larmor theorem 18
    2.7   Physical meaning of Lagrange equations; conjugate
          momenta 20
    2.8   Cyclic variables, symmetries and conserved conjugate
          momenta 23
    2.9   *Non-uniqueness of the Lagrangian 24

3   **Hamilton equations** 27
    3.1   Energy conservation 27
    3.2   Hamilton equations 30
    3.3   Coordinate transformations and Hamilton equations ... 32
    3.4   Canonical transformations 36

4   Poisson brackets and canonical structure   43
    4.1   Constants of motion identified by Poisson brackets   43
    4.2   General properties of Poisson brackets   46
    4.3   Canonical structure   48
    4.4   Invariance of the Poisson brackets under canonical
          transformations   51

5   Generation of canonical transformations   57
    5.1   Alternative characterization of canonical transformations   57
    5.2   Extended canonical transformations   65
    5.3   Generators of continuous groups of canonical
          transformations   67
    5.4   Symmetries and conservation laws. Noether theorem . . .   74
          5.4.1   Noether theorem: Lagrangian formulation   74
          5.4.2   *Noether theorem: Hamiltonian formulation   76

6   Small oscillations   83
    6.1   Equilibrium configurations. Stability   83
    6.2   Small oscillations   86

**7   *The common Poisson algebra of classical and
    quantum mechanics**   93
    7.1   Dirac Poisson algebra   93
    7.2   A common Poisson algebra of classical and quantum
          mechanics   99

8   Erratum to: Small oscillations E1

Appendix. Problems with solutions   103

Index   113

Preface

These notes were written for a half-semester introductory course of Analytical Mechanics addressed to an audience of undergraduate students, who were not theoretically and mathematically minded and in a large fraction mainly interested in experimental physics.

Now, most of the standard presentations of Analytical Mechanics (AM) derive it as a formal development of Newtonian Mechanics, devolving to the latter the discussion of the underlying physical questions. With such a choice, one risks to characterize AM essentially as a set of formal recipes, with at most the merit of mathematical elegance.

One may even be led to think that the key pillars of AM are its formal structure and the variational principles usually put at its basis, whose relevance for the discussion of concrete physical problems looks somewhat remote.

The aim of these introductory notes is to point out that, on the contrary, the basic structure of AM has very strong physical motivations, qualifying it as the simplest approach to the solutions of a large class of physical problems, with clear and complete answers.

As a matter of fact, beyond its formal aspects, the use of generalized (or Lagrangian) coordinates is an almost inevitable choice for discussing and solving problems, for which the methods of Newtonian Mechanics are not suitable and equally effective.

As discussed in these notes, the important achievements of *Lagrange equations*, with respect to Newton equations for Cartesian coordinates, are that

i) they provide the *most economical description of time evolution* in terms of the minimal set of variables necessary for describing the configurations of the system, so that there are no redundant coordinates and therefore no constraint forces;

ii) they are *form invariant* for any choice of coordinates, including those corresponding to non-inertial frames, so that they automatically take care of the problem of fictitious forces; by such an invariance, under a change of coordinates the Lagrangian transforms in the simplest possible way, i.e. by expressing in the Lagrangian function the old coordinates as functions of the new ones;

iii) finally, the full dynamical problem is completely reduced to the specification of *a single (Lagrangian) function,*(which for the large class of conservative systems is simply given by the kinetic energy minus the potential) rather than being formulated in terms of (vectorial) forces, constraints and possible terms arising in the case of non-inertial frames.

The usefulness of such properties for the solution of concrete mechanical problems is displayed by the Examples discussed in Chapters 1, 2.

A similar problem of stressing the physical motivations and effectiveness in the discussion of mechanical problems arises for the Hamiltonian formulation of Classical Mechanics.

First, the Hamiltonian has a more direct physical interpretation than the Lagrangian, being in general related to the energy function (for conservative systems the sum of the kinetic energy and the potential).

Its introduction, through the formal trick of a Legendre transform may not appeal to an experimentally minded student, if it is not motivated by the strategy of replacing the Lagrange equations, which are of second order in the time derivative, by first order equations for the canonical variables q, p.

Moreover, in this way, the initial conditions involve the initial values of the coordinates and of the conjugate momenta, which in general have a direct physical meaning (e.g., in the case of cyclic coordinates, the corresponding conjugate momenta are constants of motion).

The delicate and important issue of the relation between the Hamiltonian and the energy function, identifying the cases in which they differ, is discussed at length in Section 3.3 and in Section 5.1, with illustrating Examples.

In our opinion, the real basic pillar of the Hamiltonian formulation is the ensuing *canonical structure*, which is not only important at the conceptual level, but it also provides very useful tools for discussing and solving mechanical problems.

In particular, one gets:

a) the possibility of directly identifying the constants of motion through the vanishing of their Poisson brackets with the Hamiltonian, without having to know the solutions of the dynamical problem;

b) the *relation between symmetries of the Hamiltonian and conservation laws*;

c) the use of canonical transformations for reducing the Hamiltonian to a simpler form; the generators of infinitesimal canonical transformations with action in terms of Poisson brackets;

d) the emergence of the *canonical algebra*, etc.

The recognition of the above structures at the basis of AM, provides also a simple link with the corresponding structures in Quantum Mechanics, the emerging picture being essentially the same, once the role of the Poisson brackets is replaced by commutators.

The Chapters, Sections and Examples marked with a * were not part of the main lectures; they are not essential for the general logic of the presentation and may be skipped in a first reading or if one is not interested in more refined issues and/or developments of the basic theory.

In Section 2.9, and in Section 3.4, Example 3.13 and Remark 3.1, the problem of the non-uniqueness of the Lagrangian is discussed, completely reducing it to the addition of a total time derivative.

It is argued that such a freedom corresponds to a *gauge transformation*, since it does not change (the time evolution of) the Lagrangian variables, $q, \dot{q}$ and therefore it leaves invariant all the physical quantities, which are functions of them. The only effect is to change the (somewhat free) relation between the time derivative of the coordinates and their conjugate momenta.

This freedom has also relevant consequences in the relation between symmetries of the equations of motion and conservation laws.

By taking such a freedom into account, the Noether Theorem is revisited both in the (usual) Lagrangian formulation as well as in the (usually oversimplified) Hamiltonian formulation (Section 5.4).

As discussed in subsection 5.4.2, the invariance of the equations of motion does not require the invariance of the Hamiltonian, as usually taken for grated, but only the invariance of the Hamiltonian up to a total derivative.

The consequence is that the constant of motion related to a continuous symmetry of the dynamics is not the (canonical) generator of the infinitesimal symmetry transformation of the canonical variables, as in the oversimplified case discussed in the textbooks of Classical Mechanics, but rather the sum of the canonical generator *and* the generator of the gauge transformation corresponding to the occurrence of a total derivative ("anomalous" conservation).

The relevance of the canonical structure in Classical Mechanics and its close counterpart in Quantum Mechanics raises the conceptual question of relating it to common underlying physical principles. This issue is discussed in Chapter 7, where the Dirac proposal of a common Poisson algebra (*Dirac Poisson algebra*) is critically reviewed, and an alternative proposal is discussed, which avoids the drawbacks and mathematical inconsistencies of Dirac strategy.

The basic point of such an alternative is the realization that a common underlying physical property of both theories is the existence of translations in configuration space, leading to a Lie structure and to the Poisson algebra given by the free polynomial algebra of the coordinates and the generators of the translations. Such a *common Poisson algebra* contains a central element Z, which relates the commutator to the Lie product.

In the irreducible representations of such a common Poisson algebra Z is represented by a multiple of the identity, and the inequivalent representations are discriminated by *either* the vanishing of Z, corresponding to Classical Mechanics *or*, alternatively, by its being different from zero, and therefore *necessarily* pure imaginary, $Z = i\hbar$, corresponding to Quantum Mechanics.

The time limitation of the lectures forced a drastic limitation of the arguments to be covered and, according to the general philosophy discussed above, the general mathematical aspects have been reduced to the minimum in favor of an inductive path, starting from concrete physical problems which provide the motivations for a need of improving the Newtonian approach to Mechanics.

For these reasons, we have refrained from encumbering the notes with a long list of problems/exercises, as often done in textbooks of Classical Mechanics, with the idea that working out solutions is the best way for learning and mastering the subject.

We rather opted for a limited number of Examples/Exercises, with the task of displaying the basic ideas and mechanisms (variations on the themes being left to the student hunting).

The following inevitably incomplete presentation of AM will hopefully turn out to be useful for the students if it will convince them of the relevance and effectiveness of AM and possibly stimulate a deeper grasp of the theory, including its general mathematical structure.

For the students willing to further elaborate on AM, after this primer introduction, the following books may be useful:

L.D. Landau and E.M. Lifshitz, *Mechanics*, 3rd ed., Pergamon Press 1976,

H. Goldstein, C.P. Pool Jr., and J.L. Safko, *Classical Mechanics*, Pearson 2013,

A Fasano and S. Marmi, *Analytical Mechanics*, Oxford Graduate texts, 2006,

V.I. Arnold, *Mathematical Methods of Classical Mechanics*, Springer 1978.

1
Difficulties of Cartesian Newtonian Mechanics

1.1 Constraint forces

In this and in the following section, on the basis of simple examples, we shall discuss difficulties of the Newtonian formulation of Mechanics arising from the presence of constraint forces and fictitious forces.

For simplicity, in this section we consider a single particle of mass m; according to Newtonian Mechanics, its motion is governed and completely described by Newton equations

$$m\mathbf{a} = \mathbf{F}, \tag{1.1}$$

with $\mathbf{a}$ the acceleration and $\mathbf{F}$ the sum of the forces acting on it.

From a mathematical point of view the problem is well posed and with a unique solution (under general conditions) once the initial position and velocity are specified. However, this requires the a priori knowledge of $\mathbf{F}$, which is not the case if there are constraints.

In fact, in general eq. (1.1) may be written as

$$m\mathbf{a} = \mathbf{f} + \mathbf{R}, \tag{1.2}$$

where $\mathbf{f}$ is the sum of the (known) external forces and $\mathbf{R}$ the sum of the constraint forces, which are not a priori known, since the constraints exert just those forces necessary for constraining the motion at any

© Springer International Publishing AG 2018
F. Strocchi, *A Primer of Analytical Mechanics*, UNITEXT for Physics,
https://doi.org/10.1007/978-3-319-73761-4_1

given time. A simple example will shed light on the way of attacking the general problem.

Example 1.1. Consider the case of a simple pendulum, namely a point mass suspended from a fixed pivot through a massless inextensible rod of length r.

In this case, eq. (1.2) becomes

$$m\ddot{\mathbf{x}} = m\mathbf{g} + \mathbf{R}, \tag{1.3}$$

all vectors lying in the $x - y$ plane, with $\mathbf{g}$ the gravity acceleration and $\mathbf{R}$ the constraint force exerted by the rod (rod tension).

The constraint equation $x^2 + y^2 = r^2$ is easily solved by introducing polar coordinates, $x = r\sin\theta$, $y = r\cos\theta$. In order to cop with the unknown $\mathbf{R}$, two paths may be followed.

a) Since the constraint force does not do work (being always orthogonal to the velocity of the point mass) the energy conservation equation does not involve $\mathbf{R}$ and it is enough for solving the motion (since there is only one degree of freedom). One gets

$$\tfrac{1}{2}mr^2\dot{\theta}^2 + mgr(1 - \cos\theta) = E_0, \tag{1.4}$$

(E_0 denotes the initial energy). The rod tension, as a function of θ, may be determined by using the projection of eq. (1.3) in the direction of $\mathbf{x}/|\mathbf{x}|$, $-mr\dot{\theta}^2 = -R + mg\cos\theta$, and eq. (1.4).

b) Another possibility is to find a combination of the components of eqs. (1.2), $m\ddot{x} = -R\sin\theta$, $m\ddot{y} = -R\cos\theta + mg$, which eliminates R; in polar coordinates the result is the well known equation for the pendulum motion

$$r\ddot{\theta} = -g\sin\theta.$$

This equation is the projection of eq. (1.2) along the tangent to the trajectory (which is the standard method for the pendulum motion). It may also be obtained by taking the time derivative of eq. (1.4).

One might also describe the motion in terms of the coordinate x; for small oscillations $x \sim r\theta$, so that the above equation for θ gives

$$\ddot{x} = -(g/r)\,x. \tag{1.5}$$

The lesson from this very simple example is that in general Newton's equations in Cartesian coordinates are not the equations for the variables which describe the constrained motion.

1.2 Non-inertial frames and fictitious forces

Another source of problems for Newton's equations is the occurrence of fictitious forces in non-inertial frames, which may be more convenient frames for an intrinsic characterization of the motion of the system.

In fact, Newton's equations for coordinates relative to a non-inertial frame take the form:

$$m\mathbf{a} = \mathbf{R} + \mathbf{f}_f + \mathbf{f}, \tag{1.6}$$

where $\mathbf{R}$ denote the sum of the constraint forces, $\mathbf{f}_f$ the sum of the fictitious forces and $\mathbf{f}$ the sum of the a priori known forces.

In general, the fictitious forces are not a priori known, e.g. for a point mass the Coriolis force $2m\boldsymbol{\omega} \wedge \mathbf{v}$ (with $\boldsymbol{\omega}$ the angular velocity of the non-inertial frame) involves the velocity of the point with respect to the non-inertial frame.

Example 1.2. Consider a point mass constrained to move on a (vertical) circle of radius r, centered at the origin. The problem is to determine the motion of the point mass on the circle, when the circle rotates around the vertical diameter with angular velocity ω.

In the Cartesian coordinates x, y, z of the non-inertial reference frame in which the circle is at rest, with the x−axis orthogonal to the plane of the circle, the y-axis along the horizontal diameter and the z-axis along the vertical diameter, Newton's equations take the form

$$m\ddot{x} = R_2 - 2mv\,\omega \sin\theta, \quad m\ddot{y} = -R_1 \cos\theta + m\omega^2 r \cos\theta,$$

$$m\ddot{z} = -mg - R_1 \sin\theta, \tag{1.7}$$

where R_1 denotes the radial constraint force, R_2 the constraint force orthogonal to the plane of the circle, and the angular variable θ corresponds to polar coordinates in the $y - z$ plane, $y = r\cos\theta$, $z = r\sin\theta$. In order to determine the motion, one may start by eliminating the constraint forces. The constraint that the point mass lies on the circle implies $\ddot{x} = 0$ and $R_2 = 2mv\,\omega \sin\theta$; then, the elimination of R_1 from the two remaining equations gives

$$\ddot{z}\cos\theta - \ddot{y}\sin\theta = -g\cos\theta - \omega^2 r \sin\theta\cos\theta,$$

and in polar coordinates

$$\ddot{\theta} + (g/r)\cos\theta + \omega^2 \sin\theta\cos\theta = 0. \qquad (1.8)$$

The above Examples show that the complications of the Newtonian formulation in terms of Cartesian coordinates arise because
i) such coordinates are redundant (and in fact not independent) and *a priori unknown constraint forces* appear,
ii) the Newton's equations do not have simple transformation properties under time dependent changes of coordinates, so that *a priori unknown fictitious forces* arise in non-inertial frames.
    The following natural questions arise:
1) May one find a direct formulation of mechanical problems only in terms of the minimal number of variables necessary for describing the motion (without having to undergo somewhat laborious computations for eliminating the constraints)?
2) May one find a formulation based on equations of motion which are form-invariant under changes of coordinates?
3) May one find a formulation of mechanical problems which does not need the distinction between inertial and non-inertial frames?
    As we shall see, the Lagrangian formulation overcomes the difficulties related to the presence of constraint and fictitious forces, provides a positive answer to the above questions 1)-3) and qualifies as a very efficient way of solving dynamical problems.

2
Lagrange equations

2.1 Degrees of freedom and Lagrangian coordinates

Consider a mechanical system, whose positions are completely descri-
bed by N three-dimensional Cartesian coordinates $\mathbf{x}_1, \mathbf{x}_2, ...\mathbf{x}_N$ (for example the position of a rigid body is completely identified by three
non-collinear points). They correspond to the $3N$ Cartesian compo-
nents $x_1, y_1, z_1, x_2, y_2, z_2, ...x_N, y_N, z_N$, which, for simplicity in the following shall be denoted by x_i, $i = 1, ...3N$.

In general, when constraints are present, not all the x_i's are independent and we denote by $q_1, q_2, ...q_n$ a minimal number of (in general
non-Cartesian) "coordinates" necessary for characterizing the positions
of the (constrained) system.

The number n is called the **number of degrees of freedom** of the
system and for brevity in the following a set of minimal coordinates q_i's,
$i = 1, ...n$, will be referred to as a set of **Lagrangian coordinates**;

In contrast with the Cartesian coordinates, used in Newton's equations of motion, Lagrangian coordinates provide the most economical
description of the system and the basic question arises of directly de-
termining their time evolution without having to derive it from the
solution of the time evolution of the Cartesian coordinates.

This is the task of the Lagrangian formulation which shall be discussed below.

© Springer International Publishing AG 2018
F. Strocchi, *A Primer of Analytical Mechanics*, UNITEXT for Physics,
https://doi.org/10.1007/978-3-319-73761-4_2

If, at any time t, $\{q_i\}$ is a generic (not necessarily minimal) set of coordinates which identify the position of the system, the Cartesian coordinates x_i at time t shall be expressible as functions of the q_i's:

$$x_i(t) = x_i(q(t), t), \qquad (2.1)$$

(with q denoting the set of the q_i's) where an explicit time dependence may occur in those relations. Conversely, since the Cartesian coordinates completely determine the position of the system, they must determine the q_i's, implying that the above eqs. (2.1) are invertible:

$$q_i(t) = q_i(x(t), t).$$

In the following we shall always assume sufficient regularity to allow for the existence of the derivatives performed below. We recall that the total time derivative $\dot{x}_i$ of $x_i(t)$ gets both the contribution of the time dependence of x_i through the time dependence of q_i and the contribution from the explicit time dependence of the relations (2.1):

$$\dot{x}_i = \frac{dx_i}{dt} = \frac{\partial x_i}{\partial q_j}\frac{dq_j}{dt} + \frac{\partial x_i}{\partial t} = \frac{\partial x_i}{\partial q_j}\dot{q}_j + \frac{\partial x_i}{\partial t}, \qquad (2.2)$$

where, as often in the following, summation over repeated indices is understood.

Since $\partial x_i(q, t)/\partial q_j$ is a function of q and t, but not of $\dot{q}$, one has

$$\frac{\partial \dot{x}_i}{\partial \dot{q}_m} = \frac{\partial x_i}{\partial q_m}, \qquad (2.3)$$

a relation which will turn very useful in the following.

An explicit time dependence in the change of coordinates appears when one considers non-inertial frames.

This is simply displayed by the system discussed in Example 1.2. In fact, the position of the point on the circle is identified by the angle θ of the polar coordinates in the circle ($y = r\cos\theta$, $z = r\sin\theta$), i.e. there is only one degree of freedom, and the relation between the Cartesian coordinates x', y', z' in a fixed (inertial) frame $\mathcal{R}'$ and the Lagrangian coordinate θ is time dependent:

$$x'(t) = r\sin\theta(t)\cos(\omega t), \quad y'(t) = r\sin\theta(t)\sin(\omega t), \quad z'(t) = r\cos\theta(t),$$

where we have spelled out that the correspondence $x', y', z' \to \theta$ at time t involves the time dependence of the variables.

2.2 Lagrangian form of Newton's equations

In this section we shall show that Newton's equations in Cartesian coordinates (whose number is denoted by N) may be written in a form which we shall show to hold also for generic variables q_i, in particular for (the minimal) Lagrangian coordinates.

We consider the case of conservative forces:

$$m_i \ddot{x}_i = F_i = -\partial V(x)/\partial x_i, \quad i = 1, ... N \tag{2.4}$$

(no sum over repeated indices).

Now, denoting by T the kinetic energy $(T \equiv \frac{1}{2} \sum_i m_i \dot{x}_i^2)$, one has

$$m_i \ddot{x}_i = m_i \frac{d\dot{x}_i}{dt} = \frac{d}{dt} \frac{\partial}{\partial \dot{x}_i} \sum_j \frac{1}{2} m_j \dot{x}_j^2 = \frac{d}{dt} \frac{\partial T}{\partial \dot{x}_i}, \tag{2.5}$$

and eqs. (2.4) may be written as

$$\frac{d}{dt} \frac{\partial T}{\partial \dot{x}_i} = -\frac{\partial V}{\partial x_i}, \quad i = 1, ... N. \tag{2.6}$$

Since the potential is not a function of $\dot{x}_i$, $\partial V/\partial \dot{x}_i = 0$, and T is only a function of $\dot{x}_i$, eqs. (2.4) may also be written as

$$\frac{d}{dt} \frac{\partial L}{\partial \dot{x}_i} = \frac{\partial L}{\partial x_i}, \quad L \equiv T - V, \quad i = 1, ... N. \tag{2.7}$$

L is called the Lagrangian function or *Lagrangian* and eqs. (2.7) are the equations of motion in Lagrangian form, called the Lagrange equations in Cartesian coordinates.

Up to now, this seems nothing but a formal reshuffling of the Newton's equations, with no apparent advantage, but, as we shall see below, the substantial gain with respect to Newton equations (in Cartesian coordinates) is that the Lagrange equations hold for *any* choice of coordinates for describing the configurations of the system, with the *Lagrangian functions of different coordinates being defined by the condition of taking the same value at corresponding points.*

2.3 Lagrange equations

The Lagrange equations we are going to derive have the important property of not depending on the use of Cartesian coordinates, nor on the use of coordinates of an inertial frame, i.e. they have the form for *any* choice of coordinates q_i (*covariance of Lagrange equations under arbitrary changes of coordinates*).

As before, we consider the case of conservatives forces; we start by deriving the following equations (the needed regularity/differentiability always assumed) using eqs. (2.1), (2.2), (2.3):

$$\frac{\partial T}{\partial \dot{q}_j} = \sum_i m_i \, \dot{x}_i \frac{\partial \dot{x}_i}{\partial \dot{q}_j} = \sum_i m_i \dot{x}_i \frac{\partial x_i}{\partial q_j}, \tag{2.8}$$

$$\frac{d}{dt} \frac{\partial T}{\partial \dot{q}_j} = \sum_i (m_i \, \ddot{x}_i \frac{\partial x_i}{\partial q_j} + m_i \, \dot{x}_i \frac{d}{dt} \frac{\partial x_i}{\partial q_j}) = \sum_i F_i \frac{\partial x_i}{\partial q_j} + \frac{\partial T}{\partial q_j} \tag{2.9}$$

where, for the right hand side of eq. (2.8), we have used eq. (2.3) and for the last equality in eq. (2.9) we have used that

$$\frac{d}{dt} \frac{\partial x_i}{\partial q_j} = \frac{\partial^2 x_i}{\partial q_k \partial q_j} \dot{q}_k + \frac{\partial^2 x_i}{\partial t \, \partial q_j} = \frac{\partial}{\partial q_j} \left(\frac{\partial x_i}{\partial q_k} \dot{q}_k + \frac{\partial x_i}{\partial t} \right) = \frac{\partial \dot{x}_i}{\partial q_j}. \tag{2.10}$$

as a consequence of the allowed exchange of the order of the partial derivatives.

Introducing the components of the *generalized force*

$$Q_j \equiv \sum_i F_i \frac{\partial x_i}{\partial q_j}, \tag{2.11}$$

one has

$$\frac{d}{dt} \frac{\partial T}{\partial \dot{q}_j} = \frac{\partial T}{\partial q_j} + Q_j. \tag{2.12}$$

The generalized forces are conservatives, (i.e. they are partial derivatives of the potential with respect to the q_i's) if so are the F_i:

$$Q_j = -\sum_i \frac{\partial V}{\partial x_i} \frac{\partial x_i}{\partial q_j} = -\frac{\partial V}{\partial q_j},$$

with V independent of the $\dot{x}_j$'s, and therefore of the $\dot{q}_j$'s.

In this case, one has

$$\frac{d}{dt}\frac{\partial(T-V)}{\partial\dot{q}_j} = \frac{\partial(T-V)}{\partial q_j}.$$

Thus, one obtains the **Lagrange equations** for the *Lagrangian L*:

$$\frac{d}{dt}\frac{\partial L}{\partial\dot{q}_j} = \frac{\partial L}{\partial q_j}, \qquad L \equiv T - V, \qquad j = 1,...n. \qquad (2.13)$$

It is worthwhile to stress the crucial advantages of the Lagrange equations with respect to the Newton equations:

a) **most economical description**: the time evolution of the system may be described by the Lagrange equations for the minimal set of variables necessary for describing the system, so that there are no redundant coordinates, or constraint forces;

b) **covariance or form invariance of Lagrange equations**: the Lagrange equations they have the same form, i.e. are *form invariant*, for any choice of the coordinates, which may also be relative to noninertial frames, so that there is no problem of fictitious forces (their effect is automatically taken care of by the expression of Lagrangian as a function the time-dependent coordinates); quite generally, one of the main motivations for the Lagrangian formulation of Mechanics, is that **the Lagrangian transforms as a scalar under coordinates transformations** $q(t) \to q'(q(t), t), \ \dot{q}(t) \to \dot{q}'(q(t), \dot{q}(t), t)$, namely

$$L'(q', \dot{q}', t) = L(q(q', t), \dot{q}(q', \dot{q}', t), t); \qquad (2.14)$$

(for simplicity, in the above equation the time dependence of the coordinates has not been spelled out;

c) the formulation of the dynamical problem in terms of equations for the time evolution is reduced to the specification of a **single scalar function** (the Lagrangian) which **encodes all the relevant information**, the equations of motion being simply obtained in terms of its derivatives.

On the contrary, Newton's equations (in Cartesian coordinates) are not form invariant under coordinate transformations, in particular substantial new terms arise in the case of non-inertial frames; furthermore, the description of the motion in terms of redundant coordinates leads to the appearance of a priori unknown constraint forces.

* Remark 2.1. One may explicitly check that the presence of constraint forces due to holonomic constraints does not change eq. (2.12), with $Q_j = -\partial V/\partial q_j$, since the choice of the Lagrangian coordinates q_j, $j = 1, ...n$, with n the number of degrees of freedom, implies that the components Q_j^c of the generalized forces corresponding to the constraint forces vanish.

To this purpose, we recall that holonomic constraints on the N Cartesian coordinates which describe the position of the system are expressible in terms of conditions of the form

$$f_\alpha(x, t) = 0, \quad \alpha = 1, ...k,$$

and the constraint forces F_i^c may be written as suitable combinations of the vectors $\partial f_\alpha/\partial x_i$ orthogonal to the surfaces $f_\alpha(x, t) = 0$:

$$F_i^c = \sum_{\alpha=1}^{k} \lambda_\alpha \frac{\partial f_\alpha}{\partial x_i}, \quad i = 1, ...N.$$

Then, by introducing new variables:

$$\xi_i \equiv f_i(x, t), \quad i = 1, ...k; \quad \xi_{j+k} = q_j, \quad j = 1, ...N - k,$$

$(n = N - k$ being the number of degrees of freedom) one obtains

$$Q_j^c = \sum_i F_i^c \frac{\partial x_i}{\partial q_j} = \sum_{\alpha=1}^{k} \lambda_\alpha \sum_{i=1}^{N} \frac{\partial f_\alpha}{\partial x_i} \frac{\partial x_i}{\partial q_j} =$$

$$= \sum_{\alpha=1}^{k} \lambda_\alpha \frac{\xi_\alpha}{\partial x_i} \frac{\partial x_i}{\partial q_j} = \sum_\alpha \lambda_\alpha \frac{\partial \xi_\alpha}{\partial q_j} = 0.$$

Remark 2.2. The covariance of the Lagrange equations under a change of coordinates is implicit in the derivation of eqs. (2.13), where the q_i's are *arbitrary* (regular) functions of the Cartesian coordinates and the Lagrangian as a function of them has been defined by

$$L(q, \dot{q}, t) = L_C(x(q, t), \dot{x}(q, \dot{q}, t), t),$$

where L_C denotes the (Lagrangian) function of the Cartesian coordinates.

Thus, eq. (2.14) states that *the Lagrangians $L(q, \dot{q}, t)$ and $L_C(x, \dot{x}, t)$ take the same value when their arguments describe corresponding points.*

2.4 Lagrange equations at work. Examples

We shall check the effectiveness of the Lagrangian approach on some simple examples.

Example 1.2 As a first example, we revisit the problem, which was discussed in Section 1.2, with the Newtonian approach leading to occurrence of constraint and fictitious forces.

In the Lagrangian approach, it is convenient to use Lagrangian coordinates, i.e. a minimal set of coordinates. In this case, one may take, e.g., the angle θ, corresponding to the polar angle in the $(y - z)$-plane of the circle, and is time derivatives. Furthermore, for describing the preassigned motion of the circle we introduce the polar angle α in the $x - y$ plane.

Thus, omitting the time dependence of the coordinates x, y, z, θ spelled out in the previous discussion of Example 1.2, we have

$$x = r\cos\theta\cos\alpha, \quad y = r\cos\theta\sin\alpha, \quad z = r\sin\theta, \quad \dot{\alpha} = \omega,$$

and

$$T = \tfrac{1}{2}m\left(\dot{x}^2 + \dot{y}^2 + \dot{z}^2\right) = \tfrac{1}{2}mr^2(\dot{\alpha}^2\cos^2\theta + \dot{\theta}^2), \quad V = mgr\sin\theta. \tag{2.15}$$

The Lagrange equations immediately give eq. (1.8) without having to worry about constraint and fictitious forces.

Indeed, as argued in Remark 2.1, *the use Lagrangian coordinates eliminates the constraint forces* from the equations of motion.

Eventually, the strength of the constraint forces (needed for keeping the motion constrained, e.g. the reaction of the circle in the above Example), may be calculated at the end, once the motion has been solved, by exploiting the relation between the Lagrangian coordinates and the Cartesian coordinates (see eq. (1.7)).

An instructive issue is to understand what happens in general to the fictitious forces in the Lagrangian approach, since in the Lagrange equations only the generalized forces Q_j, which correspond to the non-fictitious forces, appear.

As we shall see below, the expression of the kinetic energy in terms of coordinates related to a non-inertial frame fully accounts for the effect of the fictitious forces. Thus, the origin of the fictitious forces is reduced to a purely kinematical ingredient, encoded in the relation between the Cartesian coordinates and the Lagrangian coordinates corresponding to a non-inertial frame.

This fact is clearly displayed by the following simple Example.

Example 2.1. Consider an inertial reference frame $\mathcal{R}$ described by Cartesian axes x, y, z and the non-inertial frame $\mathcal{R}'$ corresponding rotations of constant angular velocity ω around the z-axis.

The relation between the Cartesian coordinates of a point mass in the two frames is

$$x = x' \cos \omega t - y' \sin \omega t, \quad y = y' \cos \omega t + x' \sin \omega t, \quad z = z'$$

and a simple calculation give

$$T = \tfrac{1}{2} m \left(\dot{x}^2 + \dot{y}^2 + \dot{z}^2 \right) =$$

$$= \tfrac{1}{2} m \left(\dot{x}'^2 + \dot{y}'^2 + \dot{z}'^2 \right) + m \omega (\dot{y}' x' - \dot{x}' y') + \tfrac{1}{2} m \omega^2 (x'^2 + y'^2).$$

Thus, the expression of the kinetic energy in terms of the coordinates relative to the non-inertial frame contains additional terms, beyond the standard quadratic term in the velocities, and these terms depend also on the coordinates x', y'.

The result is that the fictitious forces arise in the Lagrange equations through the derivatives of T with respect to $x', \dot{x}', y', \dot{y}'$, a crucial role being played by the form invariance of the Lagrangian, eq. (2.14).

In fact, one has

$$\frac{\partial T}{\partial x'} = m\omega \dot{y}' + m\omega^2 x', \qquad \frac{d}{dt}\frac{\partial T}{\partial \dot{x}'} = m\ddot{x}' - m\omega \dot{y}'.$$

Thus, half the x' component of the Coriolis forces appears in the first equation, together with a part of the centrifugal force, and the other half in the second equation. Similar contributions appear in the derivative of T with respect to $y', \dot{y}'$.

By using the relation between the polar coordinates in $\mathcal{R}$ and in $\mathcal{R}'$:

$$\rho' = \rho \equiv \sqrt{x^2 + y^2 + z^2}, \qquad \theta' = \theta, \qquad \varphi' = \varphi - \omega t,$$

(with $\cos\theta \equiv z/\rho$), the kinetic energy T has the following form in terms of the polar coordinates relative to $\mathcal{R}$ and to $\mathcal{R}'$:

$$L = \tfrac{1}{2}m\left(\dot{\rho}^2 + \rho^2\,\dot{\theta}^2 + \rho^2 \sin^2\theta\,\dot{\varphi}^2\right) = \tfrac{1}{2}m\left[\dot{\rho}^2 + \rho^2\,\dot{\theta}^2 + \rho^2 \sin^2\theta(\dot{\varphi}' + \omega)^2\right].$$

Thus, in terms of the coordinates relative to the rotating frame the Lagrangian differs from the standard quadratic terms by the presence of the terms

$$L'_z\omega + \tfrac{1}{2}\omega^2\rho^2, \qquad L'_z = m\rho^2 \sin^2\theta\,\dot{\varphi}'.$$

The above expression for T in terms of the coordinates relative to the non-inertial frame $\mathcal{R}'$ may be written in vector form

$$T = \tfrac{1}{2}m\dot{\mathbf{v}}'^2 + \boldsymbol{\omega} \cdot \mathbf{L}' + \tfrac{1}{2}m\left(\boldsymbol{\omega} \wedge \mathbf{r}'\right)^2, \qquad (2.16)$$

showing that in a rotating frame the kinetic energy acquires two additional terms: i) a (vector) coupling between the angular velocity $\boldsymbol{\omega}$ of the rotating frame and the vector angular momentum $\mathbf{L}'$ in $\mathcal{R}'$, and ii) a centrifugal term.

In conclusion, the fictitious forces arise in the definition of the kinetic energy in terms of non- inertial coordinates, by exploiting the form invariance of the Lagrangian.

Example 2.2. *Eastern deviation of a falling object.*
Consider a falling particle of mass m from the height h at the equator.

The symmetry of the system suggests to use cylindrical coordinates ρ, φ, z, with ρ the distance from the center of the earth, the z-axis along the earth rotation axis, putting $z = 0$ at the equator.
In the non-inertial frame $\mathcal{R}'$ in which the earth is at rest, with coordinates given by $\rho' = \rho$, $\varphi' = \varphi - wt$, denoting by R the earth radius, one has

$$L = \tfrac{1}{2}m\left(\dot{\rho}^2 + \rho^2\,\dot{\varphi}^2\right) - mg\left(\rho - R\right) =$$
$$= \tfrac{1}{2}m\left(\dot{\rho}^2 + \rho^2(\dot{\varphi}' + w)^2\right) - mg(\rho - R). \qquad (2.17)$$

The Lagrange equations give

$$\rho^2(\dot{\varphi}' + w) = C, \quad \ddot{\rho} = -g + \rho(\dot{\varphi}' + w)^2.$$

The first equation follows from $\partial L/\partial\varphi' = 0$ and the constant C is determined by the initial conditions $\rho(0) = \rho_0$, $\varphi'(0) = 0$, i.e. $C = \rho_0^2\,w$. Then, the second equation becomes

$$\ddot{\rho} = -g + w^2\,\rho_0^4\,\rho^{-3};$$

since $w = 7,3 \times 10^{-5}$ rad/sec, $g \gg w^2\rho_0^4/\rho^3$, the second term in the above equation may be neglected to a very good approximation, yielding $\rho(t) = \rho_0 - \tfrac{1}{2}gt^2$. Hence, the equation for φ becomes

$$\dot{\varphi}' = w(\rho_0^2/\rho^2 - 1) \sim 2w(1 - \rho/\rho_0) = wg\,t^2/\rho_0$$

and it is easily solved:

$$\varphi' = \tfrac{1}{3}w\,g\,t^3/\rho_0) \sim \tfrac{1}{3}w\,g/\rho_0)(2h/g)^{3/2} > 0,$$

having used that the time t needed for reaching the earth ground is given by $t = \sqrt{2\,h/g}$.
In conclusion, the deviation from the vertical line is towards east and to a good approximation given by

$$\rho_0\varphi' = (1/3)wg(2h/g)^{3/2}.$$

Much less simple would have been the discussion in terms of Newton equations with Cartesian coordinates, as well as the control of the approximations.

Example 2.3. *Foucault pendulum.*

The problem is to determine the variation of the oscillation plane of a pendulum due to the rotation of the earth (of angular velocity $\boldsymbol{\omega}$).

In the reference frame $\mathcal{R}'$ in which the earth is at rest, it is convenient choose the z-axis along the vertical line of the pendulum and the polar coordinates r, α in the orthogonal plane tangent to the earth ground. According to eq. (2.16), neglecting ω^2 terms, as well as terms of order $1 - (z/l)^2$, with l the length of the pendulum, the kinetic energy T takes the following form

$$T = \tfrac{1}{2}m(\dot{r}^2 + r^2(\dot{\alpha} + \omega')^2),$$

with ω' the component of the angular velocity $\boldsymbol{\omega}$ along the z-axis. Furthermore, if the motion is described in terms of the coordinates of the $x - y$ plane, the gravitational force is described by a potential

$$V = -\tfrac{1}{2}(m\,g/l)r^2,$$

(see the derivation of eq. (1.5)). This may also be explained by the fact that the constraint force needed for balancing the gravitational force has a component $\mathbf{R}$ on the $x - y$ plane which is given by mgr/l , for small oscillation angles; hence, $\mathbf{R}$ is conservative and it is described by the above potential V.

Thus, the Lagrangian coincides with that of a two-dimensional harmonic oscillator written in terms of polar coordinates in a rotating frame (with angular velocity ω') and the solution is

$$r(t) = r_0(t), \quad \alpha(t) = \alpha_0(t) - \omega' t.$$

Hence, the plane of oscillations of the pendulum rotates with (constant) angular velocity ω'; at the equator $\omega' = 0$ and there is no plane rotation, whereas $\omega' = \omega$ at the north pole.

It is worthwhile to remark that more cumbersome would be the discussion in terms of Newton equations in Cartesian coordinates, with the occurrence of constraint and fictitious forces; in particular, the possibility of easily writing the Lagrangian in different coordinates has been of great help for the solution of the problem.

2.5 Generalized potential

The above derivation of the Lagrange equations under the condition of conservative forces may be generalized to the case in which there is a function $U = U(q, \dot{q}, t)$, such that the generalized forces may be written in the following form

$$Q_j \equiv \sum_i F_i \frac{\partial x_i}{\partial q_j} = \frac{d}{dt} \frac{\partial U}{\partial \dot{q}_j} - \frac{\partial U}{\partial q_j}, \tag{2.18}$$

In fact, one easily sees that the Lagrange equations hold for the Lagrangian

$$L \equiv T - U, \tag{2.19}$$

the case of conservative forces being included as a special case.

This apparently formal remark allows for the Lagrangian formulation of the motion of an *electron in the presence of an electromagnetic field* $(\mathbf{E}, \mathbf{B})$, whose corresponding forces are not derivable from a potential $V = V(x)$.

We start by showing that eq. (2.18) holds for Cartesian coordinates. In fact, the Lorentz equations for the electron position $\mathbf{x}$ and velocity $\mathbf{v}$ read

$$m \ddot{x}_i = e(E_i + \epsilon_{ijk}(v_j/c) B_k) \equiv F_i,$$

where c denotes the velocity of light, the sum over repeated induces is understood and ϵ_{ijk}, is the totally antisymmetric (Levi-Civita) tensor, $\epsilon_{123} = 1$.

In terms of the scalar and vector electromagnetic potentials $(\phi, \mathbf{A})$, one has

$$E_i = -\frac{\partial \phi}{\partial x_i} + \frac{\partial A_i}{\partial t}, \quad B_i = (\nabla \wedge \mathbf{A})_i = \epsilon_{ijk} \frac{\partial A_k}{\partial x_j},$$

Then, introducing

$$U(\mathbf{x}, \dot{\mathbf{x}}, t) \equiv e(\phi(\mathbf{x}, t) + \sum_i (\dot{x}_i/c) A_i(\mathbf{x}, t))$$

and using that

$$\frac{dA_i}{dt} = \frac{\partial A_i}{\partial t} + \dot{x}_j \frac{\partial A_i}{\partial x_j}$$

and

$$\sum_i \dot{x}_i (\nabla \wedge \mathbf{A})_i = \dot{x}_j \frac{\partial A_i}{\partial x_j} - \dot{x}_j \frac{\partial A_j}{\partial x_i},$$

one gets

$$\frac{d}{dt} \frac{\partial U}{\partial \dot{x}_j} - \frac{\partial U}{\partial x_j} = -e \frac{\partial \phi}{\partial x_i} + \frac{e}{c} \frac{dA_i}{dt} - \frac{e}{c} \dot{x}_j \frac{\partial A_j}{\partial x_i} = F_i.$$

A very important property is that eqs. (2.18) hold for any choice of coordinates.

In fact, by eq. (2.11) one has for $Q_j = F_i \, \partial x_i / \partial q_j$ (sum over the index i is understood)

$$Q_j = \left(\frac{d}{dt} \frac{\partial U}{\partial \dot{x}_i} - \frac{\partial U}{\partial x_i} \right) \frac{\partial x_i}{\partial q_j} = \frac{d}{dt} \left(\frac{\partial U}{\partial \dot{x}_i} \frac{\partial x_i}{\partial q_j} \right) - \frac{\partial U}{\partial \dot{x}_i} \frac{d}{dt} \frac{\partial x_i}{\partial q_j} - \frac{\partial U}{\partial x_i} \frac{\partial x_i}{\partial q_j} =$$

$$= \frac{d}{dt} \left(\frac{\partial U}{\partial \dot{x}_i} \frac{\partial \dot{x}_i}{\partial \dot{q}_j} \right) - \left(\frac{\partial U}{\partial \dot{x}_i} \frac{\partial \dot{x}_i}{\partial q_j} + \frac{\partial U}{\partial q_j} \right) = \frac{d}{dt} \frac{\partial U}{\partial \dot{q}_j} - \frac{\partial U}{\partial q_j}$$

where, in the third equality we have used eq. (2.3), eq. (2.10) and that the dependence of U on $\dot{q}_i$ is only through $\dot{x}_i$ (eq. (2.2)).

The possibility of writing the equations of motion of a charged particle in presence of an electromagnetic field in terms of arbitrary coordinates and the fact that the corresponding Lagrange equations are covariant under coordinate transformations greatly simplifies the changes of reference frames and consequently the treatment of electrodynamical problems.

In particular, the transformation properties of electric and magnetic fields under a change of reference frame are automatically taken care of by writing the Lagrangian, in particular the electromagnetic potential U, as a function of the new coordinates.

2.6 Larmor theorem

A useful application of the invariance of Lagrangian equations for a particle in electromagnetic field under a change of reference frame is provided by the case of an *electron bounded to a (heavy) nucleus in a constant uniform magnetic field* **H**.

Since the nucleus mass is much larger than the electron mass, in the two body problem one may neglect the nucleus motion, considering it as fixed. Then the Lagrangian for the electron motion takes the following form:

$$L = T - V_C(r) - (e/c)\mathbf{v} \cdot \mathbf{A}, \qquad (2.20)$$

with $V_C(r)$ the Coulomb potential, $A_i = \frac{1}{2}\epsilon_{ijk}x_j H_k$.

The presence of the magnetic field reduces the symmetry of the problem to a cylindrical one, and therefore it is convenient to choose the z-axis in the direction of **H** and use the following cylindrical coordinates for the electron motion

$$x = \rho\cos\theta, \quad y = \rho\sin\theta, \qquad r^2 = x^2 + y^2 + z^2 = \rho^2 + z^2.$$

Then, one has $v^2 = \dot{\rho}^2 + \rho^2\,\dot{\theta}^2 + \dot{z}^2$, $\mathbf{v}\cdot\mathbf{A} = \frac{1}{2}H\left(-\dot{x}\,y + \dot{y}\,x\right)$ and

$$L = \tfrac{1}{2}m(\dot{\rho}^2 + \rho^2\,\dot{\theta}^2 + \dot{z}^2) - V(\rho, z) - \tfrac{1}{2}(e/c)\rho^2\dot{\theta}H, \qquad (2.21)$$

where $V(\rho, z) \equiv V_C(r)$.

Since $\partial L/\partial\theta = 0$, one has $d/dt)\partial L/\partial\dot{\theta} = 0$, i.e.

$$\frac{\partial L}{\partial\dot{\theta}} = m\,\rho^2\left(\dot{\theta} - \frac{e}{2mc}H\right) = constant. \qquad (2.22)$$

Thus, the constant of motion is not the orbital angular momentum along the z-axis, $L_z = m\,\rho^2\dot{\theta}$, but rather the sum $L_z - (eH/2mc)m\rho^2$. The result is that the effect of the magnetic field amounts to adding to the orbital angular velocity $\dot{\theta}$ the angular velocity $\omega_L \equiv eH/2mc$ (**Larmor frequency**).

To better understand this fact, it is instructive to consider a reference frame $\mathcal{R}'$ rotating with velocity ω_L with respect to the previous one and the corresponding change of coordinates

$$\rho' = \rho, \quad z' = z, \quad \theta' = \theta - \omega_L t.$$

Thanks to the scalar transformation property of the Lagrangian under a change of reference frame, (eq. (2.14),

$$L'(\rho', \theta', z') = \tfrac{1}{2}m(\dot{\rho}'^2 + \rho'^2\dot{\theta}'^2 + \dot{z}'^2) - V(\rho', z') - \tfrac{1}{2}m\rho'^2\omega_L^2.$$

It is immediate to derive the equations of motion in the new frame (not so easy would be the transformation of the Newton equations, the transformation of the magnetic field requiring very special care):

$$\frac{d}{dt}\frac{\partial L}{\partial \dot{\theta}'} = \frac{\partial L}{\partial \theta'} = 0 \quad \Rightarrow \quad m\rho'^2\dot{\theta}' = \quad constant. \qquad (2.23)$$

Thus, in the rotating frame $\mathcal{R}'$, the constant of motion is the orbital angular momentum along the z-axis.

Furthermore, the effect of the magnetic field reduces to the appearance of the term $\tfrac{1}{2}m\rho'^2\omega_L^2$, which does not depend on θ' and $\dot{\theta}'$ and may therefore be absorbed in the potential, by a redefinition

$$V_C(\rho', z') \rightarrow V_C(\rho', z') + \tfrac{1}{2}m\rho'^2\omega_L^2.$$

The physical explanation is simple: the fictitious Coriolis force arising in the non-inertial reference frame $\mathcal{R}'$ counterbalances the effect of the magnetic field and the centrifugal force gives rise to the centrifugal potential $V_c = -\tfrac{1}{2}m\rho'^2\omega_L^2$, leading to a redefinition of the potential

For an atomic electron, V_c is typically very small and may be neglected to a good approximation; then, in $\mathcal{R}'$ the electron motion is the same as in the absence of the magnetic field.

The fact that the introduction of a magnetic field has the effect of changing the angular velocity even if the magnetic force has vanishing momentum (with respect to the z axis) may be explained by noting that switching on the magnetic field gives rise to a time dependent magnetic flux through the electron orbit and consequently to an electric field with non-zero momentum (with respect to the z-axis). This exactly accounts for the change of the effective angular velocity and for the change of the constant of motion.

2.7 Physical meaning of Lagrange equations; conjugate momenta

For better grasping the usefulness of Lagrangian formulation, it is helpful to discuss the physical meaning of Lagrange equations.

In the simple case of an unconstrained system subject to conservative forces described by inertial Cartesian coordinates, the Lagrange equations may be written as

$$\frac{d}{dt}p_i = \frac{\partial L}{\partial x_i}, \quad p_i \equiv \frac{\partial L}{\partial \dot{x}_i}. \tag{2.24}$$

Since in this case p_i is the momentum $m_i \dot{x}_i$ associated to the x_i coordinate, the Lagrange equations simply state that the change of p_i is governed by the force $\partial L/\partial x_i$ in the i-th direction.

In the general case, by defining the **momentum conjugated to** q_i

$$p_i \equiv \frac{\partial L}{\partial \dot{q}_i}, \tag{2.25}$$

we have an analogous interpretation: the Lagrange equations state that *the time variation of the conjugate momentum p_i is governed by the generalized force $\partial L/\partial q_j$.*

It worthwhile to remark that contributions to the generalized forces may arise also from the kinetic term T, as we have seen in the case of non-inertial Cartesian coordinates (Section 2.4).

In order to better appreciate the physical meaning of the conjugate momenta we discuss a few simple examples.

Example 2.4. Consider an electron in a central potential V.

The spherical symmetry suggests to use spherical coordinates

$$x = r\sin\theta\cos\varphi, \quad y = r\sin\theta\sin\varphi, \quad z = r\cos\theta. \tag{2.26}$$

Then, the kinetic energy reads

$$T = \tfrac{1}{2}m\left(\dot{r}^2 + r^2\dot{\theta}^2 + r^2\sin^2\theta\,\dot{\varphi}^2\right), \tag{2.27}$$

and the conjugate momenta

$$p_r = m\dot{r}, \quad p_\theta = mr^2\dot{\theta}, \quad p_\varphi = mr^2\sin^2\theta\,\dot{\varphi}, \tag{2.28}$$

have a simple physical interpretation:
1) p_r represents the radial momentum, and

$$\dot{p}_r = \partial L/\partial r;$$

2) p_φ is the z component of the angular momentum L_z, which is conserved since $\partial L/\partial\varphi = 0$ (corresponding to the vanishing of the momentum of the force relative to the z-axis).
It is convenient to choose the z-axis along the vector $\mathbf{r} = (x, y, z)$ at time zero, so that $\theta(0) = 0$ and, by the conservation equation, (omitting to spell out the time dependence of variables)

$$L_z = m\, r^2\, \dot{\varphi}\, \sin^2\theta = L_z(0) = 0.$$

If $\theta(0) = 0 = \dot{\theta}(0)$, then, using $L_z = 0$, the Lagrange equations give

$$m\ddot{r} = mr\dot{\theta}^2 - \frac{\partial V}{\partial r}, \quad \frac{d}{dt}(r^2\,\dot{\theta}) = 0,$$

which imply $r^2\dot{\theta} = r^2(0)\dot{\theta}(0) = 0$, so that $\dot{\theta}(t) = 0$ and the motion takes place along the z-axis.
Apart from this very special case, if $\sin\theta(t) \neq 0$, the equation $L_z = 0$ implies that $\dot{\varphi} = 0$, so that the motion takes place in the plane $\varphi = 0$. Then, in this case, p_θ describes the (orbital) angular momentum relative to the plane motion of the particle, and by the Lagrange equation

$$\frac{d\,p_\theta}{dt} = \frac{\partial L}{\partial\theta} = m\,r^2\sin\theta\cos\theta\,\dot{\varphi}^2 = 0,$$

p_θ is a constant of motion (*second Kepler law*).
   Less direct would be the derivation of these results by using the Newton equations in Cartesian coordinates. The advantage of the Lagrangian approach is the direct formulation of the dynamical problem in terms of the spherical coordinates, which are the more appropriate ones for a spherically symmetric problem.
Indeed, the corresponding Lagrangian (generalized) forces have a direct physical meaning, since they describe the momenta of the forces, respectively with respect to the z-axis ($M_z = \partial L/\partial\varphi$) and with respect to the (unit) normal $\mathbf{n}$ to the plane $\varphi = $ constant ($M_n = \partial L/\partial\theta$).

In fact, one has

$$\frac{\partial}{\partial\varphi} = \frac{\partial x}{\partial\varphi}\frac{\partial}{\partial x} + \frac{\partial y}{\partial\varphi}\frac{\partial}{\partial y} = -y\frac{\partial}{\partial x} + x\frac{\partial}{\partial y},$$

so that

$$\frac{\partial L}{\partial\varphi} = -y\frac{\partial L}{\partial x} + x\frac{\partial L}{\partial y} = M_z.$$

Furthermore, using that $\mathbf{n} = N\hat{\mathbf{r}} \wedge \mathbf{k}$, with $\hat{\mathbf{r}} \equiv \mathbf{r}/r$, $\mathbf{k}$ the unit vector in the direction of the z-axis and $N = \sin\theta$

$$\partial/\partial\theta = r\left(\cos\theta\cos\varphi\,\partial_x + \cos\theta\sin\varphi\,\partial_y - \sin\theta\,\partial_z\right) = -\mathbf{n}\cdot(\mathbf{r}\wedge\boldsymbol{\nabla})$$

so that $\partial L/\partial\theta = M_n$.

In conclusion, the Lagrange equations:
1) allow for the simplest description of the motion in terms of La-grangian coordinates which take into account the symmetry properties of the system;
2) directly give the dynamical laws for the angular momenta M_z, M_n, i.e. the cardinal equations of mechanics;
3) relate the conservations laws (p_φ = constant, p_θ = constant) to the symmetries of the Lagrangian;

Less transparent is the picture provided by the Newton's equation with Cartesian coordinates, where the conservation laws are not part of the equations of notion and have to be derived from them.

Example 2.3. It is useful to revisit the example of a charged particle in a Coulomb potential and in a uniform constant magnetic field, discussed in Section 2.7.

The Lagrangian (2.18) is independent of the angle θ, so that the vanishing of the generalized force $\partial L/\partial\theta$ implies the conservation of the conjugate momentum: $p_\theta = m\,\rho^2(\dot\theta - (e/2mc)H)$, which is not the orbital angular momentum of the particle, but nevertheless is a more relevant variable for describing the whole system "particle + magnetic field".

2.8 Cyclic variables, symmetries and conserved conjugate momenta

The discussion in the previous Section has shown that if the Lagrangian is independent from the Lagrangian coordinate q_i, which is then called a **cyclic variable**, the corresponding conjugate momentum $p_i = \partial L/\partial \dot{q}_i$ is a constant of motion

$$\dot{p}_i = \partial L/\partial q_i = 0. \tag{2.29}$$

As clearly displayed by the above examples, the existence of cyclic variables is related to symmetry properties of the system.

Quite generally, if q_i is a cyclic variables, the transformation

$$q_i \to q_i' = q_i + \lambda, \quad q_j \to q_j' = q_j, \quad j \neq i, \tag{2.30}$$

leaves the *Lagrangian invariant*

$$L'(q', \dot{q}', t) \equiv L(q(q', t), \dot{q}(q', \dot{q}', t), t) = L(q', \dot{q}', t),$$

and therefore corresponds to a *symmetry* of the Lagrangian and of the system.

In the Example 2.4, the cyclicity of the angle φ reflects the invariance with respect to rotations around the z-axis and implies the conservation of L_z. The angle θ becomes a cyclic variable for the Lagrangian which describes the motion in the plane $\varphi = $ constant, reflects the invariance with respect to rotations around the axis perpendicular to that plane and implies the conservation of p_θ.

In the Example 2.3 the cyclicity of the angle θ codifies the invariance under rotations around the z-axis (cylindrical symmetry).

Example 2.5. For a particle subject to a potential V, the conservation of the momentum $p_1 = m\dot{x}_1$ is equivalent to the independence of V from the Cartesian coordinate x_1, i.e. to the cyclicity of x_1 for the Lagrangian $L = T - V$. Clearly, this corresponds to the invariance of L under translations in the x_1 direction: $x_1 \to x_1 + a$.

Such a very important relation between conservation laws and invariance under coordinate transformations is clearly displayed by the Lagrangian formulation, in terms of invariance properties of L.

Example 2.6. As a further example of the relation between symmetry properties and time independence of the conjugate momenta, consider a particle constrained to move on a vertical cone with opening angle 2α subject to gravity.

The cylindrical symmetry suggests to use the distance $\rho = (x^2 + y^2)^{\frac{1}{2}}$ from the cone axis, taken as the z-axis, and the angle φ in the $x - y$ plane as Lagrangian coordinates. The cylindrical symmetry implies that the Lagrangian L is independent of φ and that the z component, L_z, of the angular momentum is a constant of motion. In fact, one has

$$L = \tfrac{1}{2}m\left(\rho^2\dot{\varphi}^2 + \frac{\dot{\rho}^2}{\sin^2\alpha}\right) - \frac{mg\rho}{\tan\alpha},$$

and $\partial L/\partial\varphi = 0$, gives $p_\varphi = m\rho^2\dot{\varphi} = L_z=$ constant. The other Lagrange equation gives

$$(d/dt)\partial L/\partial\dot{\rho} = m\ddot{\rho}/\sin^2\alpha = \partial L/\partial\rho = m\rho\dot{\varphi}^2 - mg/\tan\alpha.$$

The above equation may also be written as

$$\ddot{\rho} = \frac{L_z}{m^2\rho^3}\sin^2\alpha - g\sin\alpha\cos\alpha.$$

It is worthwhile to note that $\dot{\varphi} = L_z/m\rho^2$ is not constant in time, apart from the special case of circular orbits corresponding to $\rho = C_1$, $\dot{\varphi} = C_2$ which are solutions if the initial data satisfy $L_z(0)^2/m^2\,\rho(0)^2 = g\cot\alpha$.

2.9 *Non-uniqueness of the Lagrangian

The implicit prescription adopted so far was that the Lagrangian be defined as $L = T - V$, possibly with V replaced by the generalized potential. However, the question arises how unique is such a prescription for given equations of motions.

Clearly, a rescaling $L' = \lambda L$, $\lambda \in \mathbf{R}$ yields the same equations of motion. More generally, apart from a rescaling, the arbitrariness reduces to the addition of a total time derivative, $L' - L = dF(q)/dt$. The function F cannot depend on $\dot{q}$ because otherwise L' would involve $\ddot{q}$. (In the action integral $A = \int_{t_1}^{t_2} L(q(t), \dot{q}(t), t)\,dt$, dF/dt amounts to a *boundary term*).

In fact, the total derivative satisfies the Euler equation as an identity, *independently of the Lagrangian L*:

$$\frac{dF}{dt} = \sum_j \frac{\partial F}{\partial q_j} \dot{q}_j \quad \text{implies} \quad \frac{\partial}{\partial \dot{q}_j} \frac{dF}{dt} = \frac{\partial F}{\partial q_j},$$

and

$$\frac{d}{dt} \frac{\partial}{\partial \dot{q}_j} \frac{dF}{dt} = \frac{d}{dt} \frac{\partial F}{\partial q_j} = \sum_k \frac{\partial}{\partial q_k} \frac{\partial F}{\partial q_j} \dot{q}_k = \frac{\partial}{\partial q_j} \frac{dF}{dt}.$$

Therefore, the terms involving F disappear from the Euler equations derived from L' and one gets the same Euler equations derived from L, i.e. the same dynamical law.

The converse is also true, i.e. if $L(q, \dot{q}, t)$ and $L'(q, \dot{q}, t)$ give the same Euler equations of motion (of the form)

$$0 = \sum_j A_{ij}(q, \dot{q}, t)\ddot{q}_j + B_i(q, \dot{q}, t) =$$

$$= \frac{d}{dt} \frac{\partial L}{\partial \dot{q}_j} - \frac{\partial L}{\partial q_j} = \frac{d}{dt} \frac{\partial L'}{\partial \dot{q}_j} - \frac{\partial L'}{\partial q_j}, \qquad (2.31)$$

then (apart from a rescaling) L and L' differ by a total derivative: $L' = L + dF(q)/dt$.

In fact, putting $G \equiv L - L'$, the last equality in eq. (2.31) gives

$$\frac{d}{dt} \frac{\partial G}{\partial \dot{q}_i} - \frac{\partial G}{\partial q_i} = \frac{\partial^2 G}{\partial \dot{q}_i \partial \dot{q}_j} \ddot{q}_j + \frac{\partial^2 G}{\partial \dot{q}_i \partial q_j} \dot{q}_j + \frac{\partial^2 G}{\partial \dot{q}_i \partial t} - \frac{\partial G}{\partial q_i}.$$

Hence, by eq. (2.31) one has

$$\frac{\partial^2 G}{\partial \dot{q}_i \partial \dot{q}_j} \ddot{q}_j = 0,$$

so that G must be of the form $G = \sum_i f_i(q, t)\dot{q}_i + g(q, t)$. Then eq. (2.31) implies (sum over repeated indices understood)

$$\frac{\partial f_i}{\partial q_j} \dot{q}_j + \frac{\partial f_i}{\partial t} - \frac{\partial f_j}{\partial q_i} \dot{q}_j - \frac{\partial g}{\partial q_i} = 0.$$

Since the q's and the $\dot{q}$'s are independent Lagrangian variables, the coefficient of $\dot{q}_j$ must vanish, so that

$$\frac{\partial f_i}{\partial q_j} - \frac{\partial f_j}{\partial q_i} = 0, \quad \frac{\partial f_i}{\partial t} - \frac{\partial g}{\partial q_i} = 0. \tag{2.32}$$

By a general mathematical result (called the Poincaré Lemma) the first of eqs. (2.32) implies that (at least locally) f_i is of the form $f_i = \partial h(q, t)/\partial q_i$, and then the second of eqs. (2.32) implies

$$g(q, t) - \partial h(q, t)/\partial t = k(t).$$

In conclusion, one has

$$L - L' = \sum_i \frac{\partial h(q, t)}{\partial q_i} \dot{q}_i + \frac{\partial h(q, t)}{\partial t} + k(t) = \frac{d}{dt}\left(h(q, t) + \int_0^t d\tau\, k(\tau)\right) \tag{2.33}$$

i.e. L and L' differ by a total derivative.

Clearly, the addition of a total derivative changes the definition of the conjugate momenta:

$$p'_i \equiv \frac{\partial L'}{\partial \dot{q}_i} = p_i + \frac{\partial F(q)}{\partial q_i}.$$

3

Hamilton equations

3.1 Energy conservation

As discussed above, the Lagrange equations give the time evolution of the conjugate momenta (e.g. the momentum, the angular momentum etc.), whose conservation laws are simply related to symmetries of L. A natural question is whether one may get a similar simple derivation of the energy conservation for isolated systems in inertial frames.

We start by showing that the following function H of the coordinates q_i, $\dot{q}_i$, called the **Hamiltonian function**, is a constant of motion if the Lagrangian is not explicitly time dependent, (i.e. $\partial L/\partial t = 0$):

$$H(q, \dot{q}, t) \equiv \sum_i p_i\, \dot{q}_i - L, \quad p_i = \partial L/\partial \dot{q}_i. \tag{3.1}$$

In fact, one has (sum over i is understood)

$$\frac{dH}{dt} = \frac{d}{dt}\frac{\partial L}{\partial \dot{q}_i}\dot{q}_i + p_i\, \ddot{q}_i - \frac{\partial L}{\partial \dot{q}_i}\ddot{q}_i - \frac{\partial L}{\partial q_i}\dot{q}_i - \frac{\partial L}{\partial t} = \left(\frac{d}{dt}\frac{\partial L}{\partial \dot{q}_i} - \frac{\partial L}{\partial q_i} \right)\dot{q}_i - \frac{\partial L}{\partial t} = 0,$$

as a consequence of Lagrange equations, provided that $\partial L/\partial t = 0$. The next step is to recognize that the Hamiltonian is the total energy function, $H = T + V$, if
i) the relation between the Cartesian coordinates and the Lagrangian coordinates does not explicitly depend on time: $q_i = q_i(x)$, $i = 1, ...n$,
ii) the potential does not depend on the velocities: $\partial V/\partial \dot{q}_i = 0$, $i = 1, ...n$.

© Springer International Publishing AG 2018
F. Strocchi, *A Primer of Analytical Mechanics*, UNITEXT for Physics,
https://doi.org/10.1007/978-3-319-73761-4_3

In fact, by eq. (2.2) one has (sum over repeated indices understood)

$$T = \sum_l \tfrac{1}{2} m_l \dot{x}_l^2 = \sum_l \tfrac{1}{2} m_l \frac{\partial x_l}{\partial q_j} \frac{\partial x_l}{\partial q_i} \dot{q}_j \dot{q}_i,$$

$$p_i = \frac{\partial T}{\partial \dot{q}_i} = \sum_l m_l \frac{\partial x_l}{\partial q_j} \frac{\partial x_l}{\partial q_i} \dot{q}_j, \qquad \sum_i p_i \dot{q}_i = 2T,$$

and $H = \sum_i p_i \dot{q}_i - T + V = T + V$.

Example 3.1. Consider a point mass subject to an elastic force $k\,x$ (*harmonic oscillator*).

Then

$$L = \tfrac{1}{2} m\,\dot{x}^2 - \tfrac{1}{2} k\,x^2, \qquad p = \partial L/\partial \dot{x} = m\,\dot{x},$$

$$H = \tfrac{1}{2} m\,\dot{x}^2 + \tfrac{1}{2} k\,x^2 = p^2/2m + \tfrac{1}{2} k\,x^2 = T + V,$$

i.e. H is the energy function, whose conservation is a consequence of the fact that L is not explicitly time dependent.

Example 3.2. Particle in a central potential (see Example 2.4).

By eqs. (2.25) and (2.24), one has

$$H = m\,\dot{r}^2 + m\,r^2\,\dot{\theta}^2 + m\,r^2\,\sin^2\theta\,\dot{\varphi}^2 - L = T + V =$$

$$= \frac{p_r^2}{2m} + \frac{p_\theta^2}{2m\,r} + \frac{p_\varphi^2}{2m\,r^2\,\sin^2\theta} + V(r), \qquad (3.2)$$

i.e. the Hamiltonian function coincides with the particle energy $T + V$.

Example 3.3. Consider the case of a charged particle in presence of an electromagnetic potential (see Section 2.5).

Then, one has (sum over i understood)

$$L = \tfrac{1}{2} m\,\dot{\mathbf{x}}^2 - e(\varphi(\mathbf{x}) - (\mathbf{v}/c)\cdot\mathbf{A}(\mathbf{x})), \qquad p_i = \partial L/\partial \dot{x}_i = m\,\dot{x}_i + (e/c)\,A_i,$$

$$H = (m\,\dot{x}_i + (e/c)\,A_i)\dot{x}_i - \tfrac{1}{2} m\,\dot{x}_i\dot{x}_i + e(\varphi(\mathbf{x}) - (v_i/c)\,A_i) =$$

$$= \tfrac{1}{2} m\,\dot{\mathbf{x}}^2 + e\varphi(\mathbf{x}) = \pi_i\pi_i/2m + e\varphi(\mathbf{x}), \qquad \pi_i \equiv p_i - (e/c)\,A_i.$$

$$(3.3)$$

Thus, H is the energy function (kinetic energy + potential energy).

Example 3.4. A simple case in which the above conditions i), ii) do not hold, is provided by Example 1.2. reconsidered in Section 2.4.

$$L = \tfrac{1}{2} m r^2 (\omega^2 \cos^2 \theta + \dot{\theta}^2) - m g r \sin \theta, \quad p_\theta = m r^2 \dot{\theta}$$
$$H = \tfrac{1}{2} m r^2 \dot{\theta}^2 - L = p_\theta^2 / 2 m r^2 + m g r \sin \theta - \tfrac{1}{2} m r^2 \omega^2 \cos^2 \theta =$$
$$= T + V - m r^2 \omega^2 \cos^2 \theta,$$

(T and V given by eq. (2.15)).

Since L does not explicitly depend on time, H is a constant of motion, but it does not coincide with the particle energy, $T + V$, which is not constant in time because the system is not isolated.

In fact, work must be done for keeping the circle at constant angular velocity ω. Indeed, the work done by the external forces in the time interval dt is given by the variation of the kinetic energy

$$dW = dT = -m r^2 (\omega^2 \cos \theta \sin \theta \, \dot{\theta} + \dot{\theta} \, \ddot{\theta}) \, dt, \quad dW = -dV + dW',$$

where $dV = m g r \cos \theta \, \dot{\theta} \, dt$ is the work done by the gravitational force and dW' is the work done by the engine for keeping the angular velocity ω constant.

Then, using the equation of motion for θ, eq. (1.8), one has $dW' = 2 m r^2 \omega^2 \cos \theta \sin \theta \, \dot{\theta}$ and the work done by the engine when the point mass has moved from $\theta = -\pi/2$ to θ, is $W' = m r^2 \omega^2 \cos^2 \theta$. In conclusion, H describes the energy of the global system (particle + engine), which is constant in time.

The Lagrangian formulation allows for an easy description in the non-inertial reference frame $\mathcal{R}'$, in which the circle is at rest. Then, the particle kinetic energy in $\mathcal{R}'$ is given by $p_\theta^2 / 2 m r^2$ and the potential energy has both the contribution V due to the gravitational force and the contribution $V_c = -\tfrac{1}{2} m r^2 \omega^2 \cos^2 \theta$ due to the centrifugal force (centrifugal potential).

Thus, in such a reference frame the total energy

$$E_{\mathcal{R}'} = T_{\mathcal{R}'} + V + V_c$$

is constant in time.

Example 3.5. A similar discrepancy between the Hamiltonian function and the particle energy occurs in the case of a particle of mass m which is subject to a harmonic force $-k\mathbf{x}$, with corresponding potential $V = \frac{1}{2} k r^2$, $r \equiv |\mathbf{x}|$, and constrained to move on a horizontal rod which rotates with constant angular velocity ω around the (vertical) z-axis.

One has

$$T = \tfrac{1}{2} m(\dot{r}^2 + \omega^2 r^2), \qquad L = T - V, \qquad p_r = m\dot{r},$$

and therefore the Hamiltonian is

$$H = \tfrac{1}{2} m\dot{r}^2 - \tfrac{1}{2} m\omega^2 r^2 + V = T + V - mr^2\omega^2.$$

Since L does not explicitly dependent on time, $(\partial L/\partial t = 0)$, the Hamiltonian H is a constant of motion, whereas the energy $T + V$ is not, because $m r^2 \omega^2$ is time dependent.

As in the previous example, the constraint force $\mathbf{R}$ has a component in the direction orthogonal to the rod, $\mathbf{R} \cdot \mathbf{v} \neq 0$ and work is done by the engine which rotates the rod.

This is clearly displayed in the non-inertial frame, in which the rod is at rest, where the centrifugal force gives rise to an additional term in the potential.

3.2 Hamilton equations

The above examples indicate that in general the conjugate momenta p_i's may be more directly related to relevant physical quantities than the $\dot{q}_i$'s. In particular, the symmetry properties related to cyclic (Lagrangian) variables have a direct implication in terms of their conjugate momenta p_i.

The description of the configurations of a mechanical system in terms of the q_i's, p_i's, (called **canonical variables**), rather than in terms of the Lagrangian variables q_i's, $\dot{q}_i$'s may therefore be more convenient.

As matter of fact, in the Examples discussed above, the Hamiltonian turns out to be a function of the canonical variables. Furthermore, quite generally, the Lagrangian function $L = T - V$ does not have a direct physical meaning, in contrast with the Hamiltonian function, which, as we have seen, in many interesting cases is related to the energy, $H = T + V$.

One may then ask whether it is possible
i) to rewrite the Lagrange equations in a form which involves only the canonical variables q_i's and p_i's;
ii) to base the formulation and the equations of motion on the (more physical) Hamiltonian function, rather than on the Lagrangian function.

The positive answer to the above requirements is provided by the Hamilton equations, which, as we shall see, have also additional advantages with respect to the Lagrange equations.

To this purpose, we note that, under general conditions, the relations $p_i \equiv \partial L / \partial \dot{q}_i = p_i(q, \dot{q}, t)$ may be inverted, i.e. $\dot{q}_i = \dot{q}_i(q, p, t)$, and therefore, quite generally, the Hamiltonian function may be considered as a function,

$$H(q, p, t) \equiv \sum_i p_i \, \dot{q}_i(q, p, t) - L(q, \dot{q}(q, p, t), t)$$

of the canonical variables q, p.

Then, the differential of H may be written in terms of the differentials of two alternative sets of variables, $q_i, \dot{q}_i$ or q_i, p_i:

$$dH = \sum_i \left(\frac{\partial H}{\partial q_i} dq_i + \frac{\partial H}{\partial p_i} dp_i \right) + \frac{\partial H}{\partial t} dt,$$

$$dH = \sum_i \left(\dot{q}_i \, dp_i + p_i \, d\dot{q}_i - \frac{\partial L}{\partial \dot{q}_i} d\dot{q}_i - \frac{\partial L}{\partial q_i} dq_i \right) - \frac{\partial L}{\partial t} dt.$$

Now, comparing the two expression for dH and using $p_i \equiv \partial L / \partial \dot{q}_i$, and the Lagrange equations one gets

$$\left(\frac{\partial H}{\partial p_i} - \dot{q}_i \right) dp_i + \left(\frac{\partial H}{\partial q_i} + \dot{p}_i \right) dq_i + \left(\frac{\partial H}{\partial t} + \frac{\partial L}{\partial t} \right) dt = 0.$$

Since the variables q_i, p_i, t, and therefore their differentials, may be considered as independent, the above equation implies the **Hamilton equations**

$$\dot{q}_i = \frac{\partial H}{\partial p_i}, \qquad \dot{p}_i = -\frac{\partial H}{\partial q_i}, \qquad \frac{\partial H}{\partial t} = -\frac{\partial L}{\partial t}. \qquad (3.4)$$

Beyond the above formal manipulations, we have achieved

1) to describe the dynamics of the system by *equations which involve the (more convenient) canonical variables*;

2) to replace the Lagrange equations for the n Lagrangian coordinates, which are of second order in time and therefore require the knowledge of $2n$ independent initial data, by the *Hamilton equations* for the $2n$ independent canonical variables, which have the advantage of being *of first order in time* (the relation between $p_i(t)$ and $\dot{q}_i(t)$ is part of the Hamilton equations, not an *a priori* relation);

3) to obtain a complete description of the time evolution in terms of a *single* function, the Hamiltonian, which is (generically) related to the energy function; rather than having to specify all the forces acting on the system (some of them not a priori known), as required in the Newtonian formulation, in the Hamiltonian formulation the *whole information on the dynamical problem is encoded in the Hamiltonian* as a function of the canonical variables.

3.3 Coordinate transformations and Hamilton equations

As stressed in the previous Chapter, one of the most important and very helpful properties of the Lagrange equations is their covariance under changes of (Lagrangian) coordinates, as a consequence of eq. (2.14), and it is natural to investigate the covariance properties of the Hamilton equations, under transformations of the canonical variables.

As a first case we consider a change of coordinates, $q_i \rightarrow Q_i = Q_i(q, t)$, under which, by eq. (2.14), the Lagrangian transforms covariantly, $L_Q(Q, \dot{Q}, t) = L_q(q(Q, t), \dot{q}(Q, P, t), t)$; then, by the definition of the conjugate momenta, one has (sum over repeated indices understood)

$$P_i \equiv \frac{\partial L_Q}{\partial \dot{Q}_i} = \frac{\partial L_q(q(Q, t), \dot{q}(q, \dot{Q}, t))}{\partial \dot{Q}_i} = \frac{\partial L_q}{\partial \dot{q}_j} \frac{\partial \dot{q}_j}{\partial \dot{Q}_i} = p_j \frac{\partial \dot{q}_j}{\partial \dot{Q}_i} = p_j \frac{\partial q_j}{\partial Q_i}.$$

$$(3.5)$$

where we have used eq. (2.3), which clearly holds for any change of coordinates.

Hence, using eq. (2.2) for $\dot{q}_j$, with $q_i = q_i(Q, \dot{Q}, t)$, one has

$$\sum_i P_i \dot{Q}_i = \sum_{i,j} p_j \frac{\partial q_j}{\partial Q_i} \dot{Q}_i = \sum_j p_j \left(\dot{q}_j - \frac{\partial q_j}{\partial t} \right)$$

and for the Hamiltonian in the new canonical variables Q, P

$$H_{Q,P}(Q, P) = \sum_i P_i \dot{Q}_i - L(q(Q, t), \dot{q}(Q, P, t), t). \tag{3.6}$$

In conclusion, one has

$$H_{Q,P}(Q, P, t) = H_{q,p} - \sum_i p_i \frac{\partial q_i(Q, t)}{\partial t} \neq H_{q,p}(q(Q, t), p(Q, P, t), t).$$

$$\tag{3.7}$$

Thus, unlike the Lagrangian, the *Hamiltonian is not a scalar* under a change of coordinates. This result reflects the important physical property that in general the energy of a mechanical system changes under a change of coordinates.

We shall explicitly check this fact in some simple examples below.

Example 3.6 Consider a point mass subject to a potential V, in one dimension.

The Lagrangian and the Hamiltonian are

$$L = \tfrac{1}{2} m \dot{x}^2 - V(x), \qquad H = \tfrac{1}{2} m \dot{x}^2 + V(x) = p^2/2m + V(x),$$

and under the change of Lagrangian coordinates $x \to x' = x - v t,$, corresponding to a frame moving with a constant velocity v, one has

$$L_{x'}(x', \dot{x}', t) = L_x(x(x', t), \dot{x}(x, \dot{x}', t), t) = \tfrac{1}{2} m (\dot{x}' + v)^2 - V(x(x', t)),$$

and $p' = m(\dot{x}' + v) = p = m\dot{x}$. Then, the Hamiltonian in the original variables x, p, is $H_{x,p} = p^2/2m + V$, and

$$H_{x',p'}(x', p') = H_{x,p} - p v = \tfrac{1}{2} p'^2/2m + V(x(x', t)) - p' v.$$

showing that the Hamiltonian does not transform as a scalar under such a coordinate transformation.

Example 3.7. Consider the transformation properties of the Hamiltonian under the change of coordinates corresponding to a frame $\mathcal{R}'$ rotating with constant angular velocity ω with respect to an inertial frame $\mathcal{R}$.

For simplicity, we consider the case of a point mass subject to a potential V. In cylindrical coordinates ρ, φ, z, with the z-axis along the rotation axis of $\mathcal{R}'$, the transformation reads

$$\rho' = \rho, \ \varphi' = \varphi - \omega t, \ z' = z.$$

The Lagrangian in the new coordinates is

$$L'(\rho, \varphi', z, \dot{\rho}, \dot{\varphi}, \dot{z}) = \tfrac{1}{2}m(\dot{\rho}^2 + \rho^2(\dot{\varphi'} + \omega)^2 + \dot{z}^2) - V,$$

and $p_{\rho'} = p_\rho, \ p_{z'} = p_z, \ p_{\varphi'} = m\rho^2(\dot{\varphi'} + \omega) = p_\varphi$. Thus,

$$H(\rho, \varphi, z, p_\rho, p_\varphi, p_z) = \frac{1}{2m}\left(p_\rho^2 + p_z^2 + \frac{p_\varphi^2}{\rho^2}\right) + V(\rho, \varphi, z),$$

$$H'(\rho, \varphi', z, p_\rho, p_{\varphi'}, p_z) = \frac{1}{2m}\left(p_\rho^2 + p_z^2 + \frac{p_\varphi^2}{\rho^2}\right) + V(\rho, \varphi' + \omega t, z) - p_\varphi \omega =$$

$$= H(\rho, \varphi(\varphi', t), z, p_\rho, p_\varphi, p_z) - p_\varphi \omega, \quad p_{\varphi'} = p_\varphi.$$

Since p_φ describes the component of the angular momentum along the z-axis, the term $p_\varphi \omega$ may also be written as $\mathbf{L} \cdot \boldsymbol{\omega}$.

The Hamiltonian H is constant in time, since L is not explicitly time dependent, and describes the energy of the system in the frame $\mathcal{R}$. On the other side, H' is not constant in time, since is an explicitly time dependent function of the new variables, unless V is independent of φ; in this case $p_\varphi = L_z$ is a constant of motion and clearly so is $p_\varphi \omega$.

Example 3.8. Electron bounded to a (heavy) nucleus in a constant uniform magnetic field $\mathbf{H}$.

The Lagrangian (2.21) may be written in the form

$$L = \tfrac{1}{2}m(\dot{\rho}^2 + \dot{z}^2) + \tfrac{1}{2}m\rho^2(\dot{\theta} - \omega_L)^2 - \tfrac{1}{2}m\rho^2\omega_L^2 - V, \tag{3.8}$$

so that one easily gets

$$p_\theta = m\rho^2(\dot{\theta} - \omega_L), \quad p_\theta \dot{\theta} = p_\theta(\omega_L + p_\theta/m\rho^2)$$

and

$$H = \frac{1}{2m}\left(p_\rho^2 + p_z^2 + \frac{p_\theta^2}{\rho^2}\right) + p_\theta\,\omega_L + \tfrac{1}{2}m\,\rho^2\,\omega_L^2 + V(\rho, z). \qquad (3.9)$$

This Hamiltonian differs from the Hamiltonian of a particle subject to a potential V by the terms $p_\theta\,\omega_L$ and $\tfrac{1}{2}m\,\rho^2\,\omega_L^2$.

As we have seen in Section 2.6, the first term may be eliminated by choosing the coordinates of a rotating frame, $\theta \to \theta' = \theta - \omega_L t$. The conjugate momenta are invariant under this change of coordinates, but the Hamiltonian is not covariant $H_{\theta'} \neq H_{\theta(\theta')}$.

Example 3.9. A point mass is constrained to move on a bar of negligible mass, which forms an angle α with respect to the vertical line and rotates with angular velocity ω around it. Furthermore, the point mass is subject to an elastic force $-\tfrac{1}{2}\,k\,r$, with r the distance from the origin, and to the gravitational potential.

Clearly, the system has only one degree of freedom described by the coordinate r and one has

$$T = \tfrac{1}{2}m(\dot{r}^2 + r^2\,\omega^2\,\sin^2\alpha), \quad V = \tfrac{1}{2}k\,r^2 + m\,g\,r\cos\alpha,$$

$$L = \tfrac{1}{2}m(\dot{r}^2 + r^2\,\omega^2\,\sin^2\alpha) - \tfrac{1}{2}k\,r^2 - m\,g\,r\cos\alpha.$$

Then, the Hamiltonian is

$$H = p_r^2/2m + \tfrac{1}{2}k\,r^2 + m\,g\,\cos\alpha - \tfrac{1}{2}m\,r^2\,\omega^2\,\sin^2\alpha =$$

$$= T + V - m\,r^2\,\omega^2\,\sin^2\alpha.$$

Since the Lagrangian does not explicitly depends on time the Hamiltonian is a constant of motion, but the energy $E_\mathcal{R} = T + V$ of the particle in an inertial frame $\mathcal{R}$ is not constant in time.

On the other side, the energy $E_{\mathcal{R}'}$ with respect to the rotating frame $\mathcal{R}'$ in which the bar is at rest, coincides with H (note the presence of the centrifugal potential!) and it is constant.

3.4 Canonical transformations

Since in the Hamiltonian formulation the time evolution is described by a trajectory in the $2n$-dimensional space Γ, (called **phase space**), defined by the $2n$ independent canonical variables, it is natural to consider general invertible transformations of the coordinates of the phase space Γ, namely

$$q_i(t), p_i(t) \to Q_i(q, p, t), P_i(q, p, t). \qquad (3.10)$$

The relevant ensuing issue is then to investigate the conditions under which, *for any choice of the Hamiltonian* the corresponding Hamilton equations are (form) invariant, i.e. there is an Hamiltonian function $K(Q, P)$ such that the corresponding Hamilton equations hold for the new variables Q, P. The transformations satisfying this condition are called **canonical**.

As displayed by the above Examples, the new variables $Q_i(q, p, t)$, $P_i(q, p, t)$ cannot be assigned as arbitrary functions of the old variables.

In fact, if the new coordinates Q_i's do not depend on the p_i's, the transformation $p_i(t) \to P_i = P_i(q, p, t)$ is uniquely determined by eq. (3.5), i.e. the P_i's must be linear functions of the p_j's, with coefficients $\partial q_j / \partial Q_i$, eq. (3.5). This is the condition which guarantees that the new variables Q, P obey Hamilton equations with an Hamiltonian $H_{Q,P}$ obtained from the Lagrangian through eq. (3.6).

It is worthwhile to note that the transformation

$$q_i \to Q_i = Q_i(q), \quad p_i \to P_i = \sum_j p_j \frac{\partial q_j}{\partial Q_i} \qquad (3.11)$$

which is not explicitly dependent on time, leads to a new Hamiltonian function $H_{Q,P}$ which takes the same values as the old Hamiltonian $H_{q,p}$ for corresponding points, eq. (3.7), and this covariance property holds for any (initial) Hamiltonian function $H_{q,p}$.

For simplicity, now we consider the case of time independent transformations; the discussion of transformations which are explicitly time dependent, eq. (3.10), is more involved and it is postponed to the next Chapter.

For discussing the form invariance of the Hamilton equations it is convenient to cast them in the following compact form:

$$\dot{X_i} = G \frac{\partial H}{\partial X_i}, \quad i = 1, \dots 2n, \quad G = \begin{pmatrix} 0 & 1 \\ -1 & 0 \end{pmatrix}, \tag{3.12}$$

where X denotes a column with entries $q_1, \dots q_n, p_1, \dots p_n$ and G is a $2n \times 2n$ matrix, the entries $0, 1, 0, -1$ in eq. (3.12) being $n \times n$ matrices.

With these notations a time independent transformation of the canonical variables q_i, $p_i \rightarrow Q_i(q, p)$, $P_i(q, p)$, $i = 1, \dots n$ may be written as $X_i \rightarrow Y_i(X)$, $i = 1, \dots 2n$. Hence, one has

$$\dot{Y_i} = \sum_j \frac{\partial Y_i}{\partial X_j} \dot{X_j} = \sum_{jl} \frac{\partial Y_i}{\partial X_j} G_{jk} \frac{\partial H}{\partial Y_l} \frac{\partial Y_l}{\partial X_k}, \tag{3.13}$$

i.e. in compact matrix form

$$\dot{Y} = J G J^T \frac{\partial H}{\partial Y}, \tag{3.14}$$

where J is the Jacobian of the transformation, $(J_{ij} = \partial y_i / \partial x_j)$, and the superscript T denotes the transpose, so that

$$(J G J^T)_{il} = \sum_{jk} \frac{\partial Y_i}{\partial X_j} G_{jk} \frac{\partial Y_l}{\partial X_k}.$$

Thus, the transformation leaves the Hamilton equations invariant, for *any* chosen Hamiltonian H, if and only if

$$J G J^T = G. \tag{3.15}$$

This condition characterizes a **canonical transformation**.

As we shall see later, this condition characterizes also the time dependent canonical transformations; in this more general case the new Hamiltonian K differs from the old one H by a Hamiltonian-independent term, which does only depend on the canonical transformation (see eqs. (3.7) and Examples 3.6-3.9).

It is easy to check that the condition is satisfied by the transformation (3.11).

Example 3.10. It is easy to check that the following transformations are canonical:

a) $Q_i = p_i, \quad P_i = -q_i$

b) $Q_i = \lambda q_i, \quad P_i = \mu q_i, \quad \lambda \mu = 1$. The condition $\lambda \mu = 1$ is necessary and sufficient for the transformation being canonical.

The canonicity of the transformations b), may be used to transform the Hamiltonian of the harmonic oscillator in the following simple form:

$$H = \tfrac{1}{2}(p^2/m + k^2 q^2) \to H = \tfrac{1}{2}w(P^2 + Q^2), \quad w \equiv k/m,$$

by taking $\lambda = \sqrt{m\,w}, \mu = \lambda^{-1}$.

Example 3.11. Verify that the transformation:

$$Q_i = \alpha\, q_i + \beta\, p_i, \quad P_i = \gamma\, q_i + \delta\, p_i,$$

is canonical if and only if $\alpha\,\delta - \beta\,\gamma = 1$.

*Example 3.12. A transformation

$$q \to Q(q, t), \quad p \to P(q, p, t)$$

satisfies condition (3.15) if and only if $P(q, p, t)$ is of the form

$$P_i = \sum_j p_j \frac{\partial q_j}{\partial Q_i} + \frac{\partial G(q)}{\partial q_i}.$$

We have to exploit the condition given eq. (3.15). To this purpose, we note that J has the form

$$J = \begin{pmatrix} \partial Q & 0 \\ \partial_q P & \partial_p P \end{pmatrix},$$

where

$$(\partial Q)_{ij} \equiv \frac{\partial Q_i}{\partial q_j}, \quad (\partial_q P)_{ij} \equiv \frac{\partial P_i}{\partial q_j}, \quad (\partial_p P)_{ij} \equiv \frac{\partial P_i}{\partial p_j}.$$

Then, eq. (3.15) reads

$$JGJ^T = \begin{pmatrix} \partial Q & 0 \\ \partial_q P & \partial_p P \end{pmatrix} \begin{pmatrix} 0 & 1 \\ -1 & 0 \end{pmatrix} \begin{pmatrix} (\partial Q)^T & (\partial_q P)^T \\ 0 & (\partial_p P)^T \end{pmatrix} =$$

$$\begin{pmatrix} 0 & \partial Q\, (\partial_p P)^T \\ -\partial_p P\, (\partial Q)^T & -\partial_p P\, (\partial_q P)^T + \partial_q P\, (\partial_p P)^T \end{pmatrix} = \begin{pmatrix} 0 & 1 \\ -1 & 0 \end{pmatrix}.$$

First, this condition implies $\partial_p P = (\partial Q^T)^{-1}$, so that $\partial_p P$ must be independent of p, since so is ∂Q; moreover, by definition

$$(\partial Q^T)_{ij} = \partial Q_j / \partial q_i, \quad (\partial Q^T)_{ij}^{-1} = \partial q_j / \partial Q_i$$

and therefore one has

$$P_i = \sum_j p_j\, (\partial Q^T)_{ij}^{-1} + G_i(q) = \sum_j p_j\, \partial q_j / \partial Q_i + G_i(q).$$

Finally, $-\partial_p P\, (\partial_q P)^T + \partial_q P\, (\partial_p P)^T = 0$ requires that

$$\frac{\partial G_j}{\partial Q_i} - \frac{\partial G_i}{\partial Q_j} = 0$$

and by Poincaré Lemma G_i is of the form of a "gradient", $G_i = \partial G / \partial Q_i$ for some function G, equivalently $G_i = \partial G / \partial q_i$.

*Example 3.13 The addition of a total derivative to the Lagrangian

$$L \to L + dG/dt \equiv L'$$

changes the definition of the conjugate momenta but it gives rise to equivalent Hamilton equations, i.e. the same dynamics is described in terms of different canonical variables.

In fact, from L' one gets

$$p'_j \equiv \frac{\partial L'}{\partial \dot{q}_j} = p_j + \frac{\partial G(q)}{\partial q_j}, \quad \frac{\partial p_k}{\partial p'_j}\Big|_q = \delta_{kj},$$

where $|_q$ indicates that q has to be kept fixed in performing the partial derivative with respect to the other variables.

Furthermore, by eqs. (5.30), $q_i(t)$ and $\dot{q}_i(t)$ do not change. Then, (sum over repeated indices understood)

$$H'(q, p') = p'_j \dot{q}_j - L' = (p_j + \frac{\partial G}{\partial q_j}) \dot{q}_j - L - \frac{\partial G}{\partial q_j} \dot{q}_j = H(q, p) =$$

$$= H(q, p' - \frac{\partial G(q)}{\partial q}). \tag{3.16}$$

The corresponding Hamilton equations are

$$\frac{dq_j}{dt} = \frac{\partial H'}{\partial p'_j}\Big|_q = \frac{\partial H}{\partial p_k} \frac{\partial p_k}{\partial p'_j}\Big|_q = \frac{\partial H}{\partial p_j}\Big|_q,$$

$$\frac{dp'_j}{dt} = -\frac{\partial H'}{\partial q_j}\Big|_{p'} = -\frac{\partial H}{\partial q_j}\Big|_p - \frac{\partial H}{\partial p_k} \frac{\partial p_k}{\partial q_j}\Big|_{p'} = -\frac{\partial H}{\partial q_j}\Big|_p + \dot{q}_k \frac{\partial}{\partial q_j} \frac{\partial F}{\partial q_k}.$$

If, in the right hand side of the last equation p' is replaced by its expression in terms of p, the terms involving G cancel and one reobtains the equation for $\dot{p}$ corresponding to H.

* **Remark 3.1** The transformation

$$Q_i(t) = q_i(t), \quad \dot{Q}_i(t) = \dot{q}_i(t), \quad P_i(t) = p_i(t) + \partial G(q)/\partial q_i \tag{3.17}$$

is a special case of the transformations discussed in Example 3.12, and therefore is canonical.

It does not change the Lagrangian variables $q, \dot{q}$ and therefore it leaves all the physical quantities $F(q, \dot{q}, t)$ invariant. However, it changes the relation between the canonical momentum and $\dot{q}$, corresponding to the addition of a total derivative $d\,G(q)/d\,t$ to the Lagrangian.

For these reasons, one may call the transformation (3.17) a *"gauge" transformation*.

As a matter of fact, in the simple case of a particle in the presence of a magnetic field, see Example 3.3, the particle position $\mathbf{x}$ and velocity $\mathbf{v} = \dot{\mathbf{x}}$ are observable quantities, but the canonical momentum $p_i = \dot{x}_i + (e/c)A_i$ is not invariant under a gauge transformation $A_i \to A_i + \partial \Lambda(\mathbf{x})/\partial x_i$. In fact, one has

$$x_i \to x_i, \quad \dot{x}_i \to \dot{x}_i, \quad p_i \to p_i + (e/c)\partial \Lambda/\partial x_i,$$

corresponding to the fact that under a gauge transformation the Lagrangian is invariant up to a total derivative.

Under the canonical transformation (3.17), the Hamiltonian is invariant up to a total derivative of a function of the q's.

In fact, by eq. (3.16), one has:

$$H'(q, p') = H(q, p)$$

and, on the other side,

$$H(q, p') = \dot{q}_i p_i + \dot{q}_i \frac{\partial G(q)}{\partial q_i} - L =$$

$$= H(q, p) + \frac{dG(q)}{dt}.$$

Hence, by comparing the two equations, one has

$$H'(q, p') = H(q, p') - \frac{dG(q)}{dt}, \tag{3.18}$$

i.e. the Hamiltonian is invariant up to a total derivative.

4

Poisson brackets and canonical structure

4.1 Constants of motion identified by Poisson brackets

Since the canonical variables completely describe the state of the mechanical system, any physical quantity F shall be described by a function of the canonical variables: $F(q, p, t)$, possibly with an explicit time dependence.

A very relevant issue for the discussion of the dynamics is whether a physical quantity $F(q(t), p(t), t)$ is a constant of motion. In principle, one should know the solution $q(t), p(t)$ of the evolution equations and then check if, correspondingly

$$\frac{dF}{dt} = \sum_i \left(\frac{\partial F}{\partial q_i} \dot{q}_i + \frac{\partial F}{\partial p_i} \dot{p}_i \right) + \frac{\partial F}{\partial t} = 0. \tag{4.1}$$

Clearly, such a procedure is not of great help since it requires the full knowledge of the dynamics, which the constants of motion should be instrumental to provide helpful information about.

As we have seen above, the occurrence of cyclic variables (typically corresponding to symmetries of the Lagrangian or of the Hamiltonian) provide immediate information, but in general one faces the problem of checking eq. (4.1). As we shall see, the Hamiltonian formulation turns out to be very useful, since it provides an easy solution of this problem.

© Springer International Publishing AG 2018
F. Strocchi, *A Primer of Analytical Mechanics*, UNITEXT for Physics,
https://doi.org/10.1007/978-3-319-73761-4_4

In fact, by using the Hamilton equations, eq. (4.1) becomes

$$\frac{dF}{dt} = \sum_i \left(\frac{\partial F}{\partial q_i} \frac{\partial H}{\partial p_i} - \frac{\partial F}{\partial p_i} \frac{\partial H}{\partial q_i} \right) + \frac{\partial F}{\partial t} = 0 \qquad (4.2)$$

and may be checked *without having to know the solution* $q_i(t)$, $p_i(t)$, only the expression of the Hamiltonian function in terms of the canonical variables being sufficient.

No similar simple information is provided by the Lagrangian, apart from the case of cyclic variables, (not to speak of the Newtonian approach in Cartesian coordinates) and this adds further support to the useful role of the Hamiltonian function.

As displayed on many occasions below, given two physical quantities $F(q, p, t)$, $G(q, p, t)$ a convenient concept is their **Poisson bracket**

$$\{F, G\} \equiv \sum_i \left(\frac{\partial F}{\partial q_i} \frac{\partial G}{\partial p_i} - \frac{\partial F}{\partial p_i} \frac{\partial G}{\partial q_i} \right). \qquad (4.3)$$

With such a notation, the time derivative of F takes the form

$$\frac{dF}{dt} = \{F, H\} + \frac{\partial F}{\partial t}, \qquad (4.4)$$

and therefore, if F is not explicitly time dependent, its being a constant of motion is equivalent to the vanishing of its Poisson bracket with the Hamiltonian.

Examples The power of the Poisson brackets for checking whether a physical quantity $F(q, p, t)$ is a constant of motion (without having to solve the equations of motion) may be seen by revisiting the Examples discussed before.

In the Example 3.4, it is immediate to check that $\{T + V, H\} \neq 0$, and therefore the energy $T + V$, in the inertial reference frame $\mathcal{R}$, is not a constant of motion) whereas so is $T_{\mathcal{R}'} + V + V_c$.

Similarly, in Examples 3.5, 3.9, $\{T + V, H\} \neq 0$, so that again $T + V$ is not a constant of motion.

Example 3.7. (See Section 3.3) By using the Poisson brackets, it is immediate to check that the energy in the reference frame $\mathcal{R}$ is a constant of motion, but the Hamiltonian H' in the non inertial reference frame $\mathcal{R}'$,

$$H' = \frac{1}{2m} \left(p_\rho^2 + p_z^2 + \frac{p_\theta^2}{\rho^2} \right) + V(\rho, \varphi' - \omega t, z)$$

is not a constant of motion:

$$\frac{dH'}{dt} = \{H', H'\} + \frac{\partial H'}{\partial t} = \frac{\partial V}{\partial t} = \frac{\partial V}{\partial \varphi} \omega = -\dot{p}_\varphi \omega,$$

unless V is independent of φ.

Example 4.1. Consider two particles with an interaction described by a potential $V(\mathbf{x}_1, \mathbf{x}_2)$. Are there constants of motion of the form $F(\mathbf{x}_1, \mathbf{x}_2)$, $F(\mathbf{p}_1, \mathbf{p}_2)$?

In terms of the Poisson brackets, the condition that $F(\mathbf{x}_1, \mathbf{x}_2)$ is a constant of motion reads

$$\sum_i \left(\frac{\partial F}{\partial x_{1,i}} p_{1,i} + \frac{\partial F}{\partial x_{2,i}} p_{2,i} \right) = 0.$$

Since, in the initial conditions, the positions and the momenta may be assigned independently, the above equation requires $\partial F / \partial x_{1,i} = 0 = \partial F / \partial x_{2,i}$, i.e only the constant functions are allowed.
On the other side, the condition that $F(\mathbf{p}_1, \mathbf{p}_2)$ is a constant of motion reads

$$\sum_i \left(\frac{\partial F}{\partial p_{1,i}} \frac{\partial V}{\partial x_{1,i}} + \frac{\partial F}{\partial p_{2,i}} \frac{\partial V}{\partial x_{2,i}} \right) = 0.$$

This condition is easily solved if $V(\mathbf{x}_1, \mathbf{x}_2) = V(\mathbf{x}_1 - \mathbf{x}_2)$, since then $\partial V / \partial x_{1,i} = \nabla_i V = -\partial V / \partial x_{2,i}$, and the condition requires

$$\frac{\partial F}{\partial p_{1,i}} - \frac{\partial F}{\partial p_{2,i}} = 0.$$

Thus, the only functions allowed are all the functions of $\mathbf{P} = \mathbf{p}_1 + \mathbf{p}_2$.

4.2 General properties of Poisson brackets

The Poisson brackets define a very important algebraic structure on the algebra of regular functions of the canonical coordinates, as discussed below.

To avoid technical points we shall consider the algebra $\mathcal{A}$ of the infinitely differentiable (briefly C^∞) real functions of the canonical variables (including the possibility of a differentiable explicit dependence on time, $F(q, p, t)$, even if, for simplicity it shall not always spelled out in the following).

We start by recalling that the *standard* algebraic operations in $\mathcal{A}$ are defined by :

i) linear (vector space) structure

$$(\lambda F)(q, p) \equiv \lambda F(q, p), \quad \forall \lambda \in \mathbf{R}; \quad (F + G)(q, p) \equiv F(q, p) + G(q, p);$$

ii) algebraic product

$$(F\,G)(q, p) \equiv F(q, p)\,G(q, p), \quad F, G \in \mathcal{A}.$$

The so defined product is clearly associative:

$$(F\,(G\,K)) = ((F\,G)\,K).$$

and therefore $\mathcal{A}$ is an *associative algebra*.

An additional "product" is defined by the Poisson brackets: $A, B \to \{A, B\} \in \mathcal{A}$,

$$\{A, B\} \equiv \sum_i \left(\frac{\partial A}{\partial q_i} \frac{\partial B}{\partial p_i} - \frac{\partial A}{\partial p_i} \frac{\partial B}{\partial q_i} \right), \quad \forall A, B \in \mathcal{A}. \qquad (4.5)$$

The Poisson brackets satisfy the following general properties, which easily follow from the definition and turn out to be very useful for the computations:

1) (*antisymmetry*)
$$\{A, B\} = -\{B, A\}, \qquad (4.6)$$

2) (*linearity in both factors*)
$$\{A, B + C\} = \{A, B\} + \{A, C\}, \qquad (4.7)$$

3) (*Leibniz rule*)

$$\{A, \, BC\} = \{A, \, B\}C + B\{A, \, C\}, \tag{4.8}$$

4) (*Jacoby identity*)

$$\{A, \, \{B, \, C\}\} + \{C, \, \{A, \, B\}\} + \{B, \, \{C, \, A\}\} = 0. \tag{4.9}$$

The Jacoby identity states that the sum of the trilinear Poisson brackets obtained by cyclically permuting the three variables A, B, C vanishes.

The above properties characterize basic mathematical structures. Given an associative algebra a bilinear map satisfying properties 1, 2, 4, is called a Lie bracket and the additional property 3 (Leibniz rule) states that it is a derivation. A prototypic example is provided by a Lie algebra.

In the following, a bilinear map satisfying 1-4 shall be called a **Lie product**. Technically, a real associative algebra $\mathcal{A}$, (i.e. with an associative product), equipped with a Lie product is called a **Poisson Algebra**.

The properties 1)-4) imply the following useful result (*Jacobi theorem*:

The Poisson bracket of two constant of motion is also a constant of motion.

In fact, by using the Jacoby identity, one has:

$$\frac{d\{A, \, B\}}{dt} = \{\{A, \, B\}, \, H\} + \frac{\partial\{A, \, B\}}{\partial t} =$$

$$= -\{\{H, \, A\}, B\}\} - \{\{B, \, H\}, A\} + \{\frac{\partial A}{\partial t}, \, B\} + \{A, \, \frac{\partial B}{\partial t}\} =$$

$$= \{\frac{dA}{dt}, \, B\} + \{A, \, \frac{dB}{dt}\} = 0. \tag{4.10}$$

The result is not as trivial as it might appear, the Leibniz rule applies to the time derivative acting on the Poisson brackets thanks to the Jacobi identity.

4.3 Canonical structure

For the applications is it useful to consider the Poisson brackets of some basic physical quantities.

First, we consider the canonical variables q, p; from the definition, eq. (4.3), one easily gets:

$$\{q_i, q_j\} = 0, \quad \{p_i, p_j\} = 0, \quad \{q_i, p_j\} = \delta_{ij}, \tag{4.11}$$

where δ_{ij} is the Kronecker symbol, $\delta_{ij} = 1$ for $i = j$, and zero otherwise.

The above Poisson brackets are called the *canonical Poisson brackets*. For the polynomial algebra of canonical variables, $\mathcal{A}_P$, they encode all the information given by the explicit definition of the Poisson brackets; in fact, they may be used to define a unique Lie product satisfying 1-4, which actually coincides with the Poisson bracket explicitly defined by eq. (4.3).

By the definition of Poisson brackets, one also has

$$\{q_i, B(q, p, t)\} = \frac{\partial B}{\partial p_i}, \quad \{p_i, B(q, p, t)\} = -\frac{\partial B}{\partial q_i}. \tag{4.12}$$

For the polynomial algebra $\mathcal{A}_P$ they are implies by the canonical Poisson brackets (4.11). It may be useful to remark that eqs. (4.11), (4.12) have a strict counterpart in Quantum Mechanics.

Next, we consider the Poisson brackets of the angular momentum $L_i = \varepsilon_{ijk} x_j p_k$; one gets (the sum over repeated indices is understood)

$$\{x_i, L_j\} = \epsilon_{ijk} x_k, \quad \{p_i, L_j\} = \epsilon_{ijk} p_k, \tag{4.13}$$

$$\{L_i, L_j\} = \epsilon_{ijk} L_k, \quad \{L_1, L_2\} = L_3, \{L_1, L_3\} = -L_2, \{L_2, L_3\} = L_1, \tag{4.14}$$

$$\{\mathbf{L}^2, L_i\} = L_j \{L_j, L_i\} + \{L_j, L_i\} L_j = \epsilon_{jik} (L_j L_k + L_k L_j) = 0. \tag{4.15}$$

The following Examples illustrate the usefulness of the above canonical structure, properties 1-4 and eqs. (4.11-15).

Example 4.2. If two components of the angular momentum are constants of motion, so is also the third one.

In fact, if L_1, L_2 are constants of motion, by Jacobi theorem, (eq. (4.10)), also $L_3 = \{L_1, L_2\}$ is a constant of motion:

Example 4.3. Consider two harmonic oscillators with a mutual inter-
action described by a potential $V(|\mathbf{x}^{(1)} - \mathbf{x}^{(2)}|)$. Then, the Hamiltonian
is

$$H = \tfrac{1}{2}(\mathbf{p}_{(1)}^2 + \mathbf{p}_{(2)}^2)/m + \tfrac{1}{2}(k_{(1)}^2\,\mathbf{x}_{(1)}^2 + k_{(2)}^2\,\mathbf{x}_{(2)}^2) + V(|\mathbf{x}^{(1)} - \mathbf{x}^{(2)}|).$$

What may be said about the constants of motion?

As before, no $F(\mathbf{x}^{(1)}, \mathbf{x}^{(2)})$ is a constant of motion. By exploiting the
independence of the initial conditions for the positions and the mo-
menta and by arguing as before one concludes that there is no constant
of motion of the form $F(\mathbf{p}^{(1)}, \mathbf{p}^{(2)})$.
Interesting functions are the angular momenta of the two oscillators
$\mathbf{L}^{(1)}$, $\mathbf{L}^{(2)}$. By using the above eqs.(4. 11-13) one gets

$$\{L_i^{(1)}, H\} = \epsilon_{ijk}\, p_k^{(1)}\, p_j^{(1)}/m - \epsilon_{ijk}\, x_k^{(1)}\, (k_{(1)}^2\, x_j^{(1)} + \frac{\partial V}{\partial x_j^{(1)}}) =$$

$$= V'\,\epsilon_{ijk}\, \frac{x_k^{(1)}\, x_j^{(2)}}{|\mathbf{x}^{(1)} - \mathbf{x}^{(2)}|} \neq 0, \quad \{L_i^{(2)}, H\} = V'\,\epsilon_{ijk}\, \frac{x_k^{(2)}\, x_j^{1}}{|\mathbf{x}^{(1)} - \mathbf{x}^{(2)}|} \neq 0,$$

and $\{L_i^{(1)} + L_i^{(2)}, H\} = 0$.

***Example 4.4.** (Spherical and cylindrical canonical variables)*
It is an instructive exercise to check that the transformation from Cartesian to spherical or cylindrical canonical variables is canonical.

Spherical canonical variables.
It is convenient to write the transformation in matrix form, with the
coordinates written in a column:

$$\mathbf{x} = \begin{pmatrix} x \\ y \\ z \end{pmatrix} = r \begin{pmatrix} \sin\theta\cos\varphi \\ \sin\theta\sin\varphi \\ \cos\theta \end{pmatrix}. \tag{4.16}$$

Then, one has

$$\mathbf{v} = \dot{r}\begin{pmatrix} \cos\theta\cos\varphi \\ \cos\theta\sin\varphi \\ -\cos\theta \end{pmatrix} + r\dot{\theta}\begin{pmatrix} \sin\theta\cos\varphi \\ \sin\theta\sin\varphi \\ \cos\theta \end{pmatrix} + r\sin\theta\dot{\varphi}\begin{pmatrix} -\sin\varphi \\ \cos\varphi \\ 0 \end{pmatrix}.$$

Hence, (the superscript T denotes the transpose in the matrix multiplication)

$$T = \tfrac{1}{2}m\,\mathbf{v}\cdot\mathbf{v} = \tfrac{1}{2}m\,\mathbf{v}^T\mathbf{v} = \tfrac{1}{2}m\,(\dot{r}^2 + (r\,\dot\theta)^2 + (r\sin\theta\,\dot\varphi)^2);\qquad (4.17)$$

$$p_r = m\,\dot r,\; p_\theta = m\,r^2\,\dot\theta^2,\; p_\varphi = m\,r^2\,\sin^2\theta\,\dot\varphi,$$

$$T = \frac{1}{2m}p_r^2 + \frac{1}{2mr^2}\left(p_\theta^2 + \frac{p_\varphi^2}{\sin^2\theta}\right).\qquad (4.18)$$

It is useful to express the angular momentum $\mathbf{L} = m\,\mathbf{x}\wedge\mathbf{v}$ in spherical coordinates:

$$\mathbf{L} = m\,r^2\left[\dot\theta\begin{pmatrix}-sin\varphi\\ \cos\varphi\\ 0\end{pmatrix} - \sin\theta\,\dot\varphi\begin{pmatrix}\cos\theta\cos\varphi\\ \cos\theta\sin\varphi\\ -\sin\theta\end{pmatrix}\right] =$$

$$= p_\theta\begin{pmatrix}-\sin\varphi\\ \cos\varphi\\ 0\end{pmatrix} - \frac{p_\varphi}{\sin\theta}\begin{pmatrix}\cos\theta\cos\varphi\\ \cos\theta\sin\varphi\\ -\sin\theta\end{pmatrix},\qquad (4.19)$$

in particular $L_z = p_\varphi$.

Clearly $\{L_z, T\} = 0$, a result already obtained before. Actually, in addition one has $\{L_x, T\} = 0$, and therefore, by Jacobi theorem, also $\{L_y, T\} = 0$.

Cylindrical canonical variables.

As before, it is convenient to write the transformation from Cartesian to cylindrical coordinates in matrix form:

$$\mathbf{x} = \begin{pmatrix}x\\ y\\ z\end{pmatrix} = \begin{pmatrix}\rho\cos\varphi\\ \rho\sin\varphi\\ z\end{pmatrix};\qquad (4.20)$$

$$\mathbf{v} = \dot\rho\begin{pmatrix}\cos\varphi\\ \sin\varphi\\ 0\end{pmatrix} + \rho\dot\varphi\begin{pmatrix}-\sin\varphi\\ \cos\varphi\\ 0\end{pmatrix} + \dot z\begin{pmatrix}0\\ 0\\ 1\end{pmatrix}.$$

Then,

$$T = \tfrac{1}{2} m(\dot{\rho}^2 + (\rho\,\dot{\varphi})^2 + \dot{z}^2) = \frac{1}{2m}\left(p_\rho^2 + \frac{p_\varphi^2}{\rho^2} + p_z^2\right),$$

$p_\rho = m\,\dot{\rho}, \quad p_\varphi = m\,\rho^2\,\dot{\varphi}, \quad p_z = m\,\dot{z}.$

For the angular momentum in cylindrical coordinates, one has

$$\mathbf{L} = \begin{pmatrix} \rho\sin\varphi\,p_z - z\sin\varphi\,p_\rho - z\cos\varphi\,p_\varphi/\rho \\ -\rho\cos\varphi\,p_z + z\cos\varphi\,p_\rho - z\sin\varphi\,p_\varphi/\rho \\ p_\varphi \end{pmatrix}.$$

Clearly, the transformation: $\mathbf{x}, \mathbf{p} \to (r,\,\theta,\,\varphi), (p_r, p_\theta, p_\varphi)$ defined above, leading from Cartesian to spherical canonical variables, satisfies eq. (3.5) (by the way the canonical momenta have been defined) and therefore it is canonical.

The same property holds for the transformation which leads from Cartesian to cylindrical canonical coordinates: $\mathbf{x}, \mathbf{p} \to (\rho, \varphi, z), (p_\rho, p_\varphi, p_z)$.

This implies that both the spherical and the cylindrical canonical coordinates satisfy eqs. (4.11), (4.12).

4.4 Invariance of the Poisson brackets under canonical transformations

Given a canonical transformation $q, p \to Q(q, p, t), P(q, p, t)$, for any two physical quantities A, B one may define the Poisson brackets by using the two different canonical variables

$$\{A,\,B\}_{q,p} = \sum_i \left(\frac{\partial A}{\partial q_i}\frac{\partial B}{\partial p_i} - \frac{\partial A}{\partial p_i}\frac{\partial B}{\partial q_i}\right), \qquad (4.21)$$

$$\{A,\,B\}_{Q,P} = \sum_i \left(\frac{\partial A}{\partial Q_i}\frac{\partial B}{\partial P_i} - \frac{\partial A}{\partial P_i}\frac{\partial B}{\partial P_i}\right), \qquad (4.22)$$

where, in eqs. (4.22), as functions of the new variables A and B are defined by $A(q(Q, P, t), p(Q, P, t), t), B(q(Q, P, t), p(Q, P, t), t)$.

A very important and useful property is that the two definitions coincide (**invariance of the Poisson brackets under canonical transformations**).

In fact, by introducing the vectors

$$x = (x_1 = q_1, ..., x_n = q_n, x_{n+1} = p_1, ..., x_{2n} = p_n),$$

$$y = (y_1 = Q_1, ..., y_n = Q_n, y_{n+1} = P_1, ..., y_{2n} = P_n),$$

(see Section 3.4), and denoting by J the Jacobian of the transformation, one has (sum over k understood) $\partial A / \partial x_i = \partial A / \partial y_k \, J_{ki}$. Hence

$$\{A, B\}_{q,p} = \sum_{i=1}^{n} \left(\frac{\partial A}{\partial q_i} \frac{\partial B}{\partial p_i} - \frac{\partial A}{\partial p_i} \frac{\partial B}{\partial q_i} \right) = \sum_{i,j=1}^{2n} \left(\frac{\partial A}{\partial x_i} G_{ij} \frac{\partial B}{\partial x_j} \right), \quad (4.23)$$

with G defined by eq. (3.12), and thanks to eq. (3.15) (with the sum over repeated indices understood)

$$\{A, B\}_{q,p} = \frac{\partial A}{\partial y_i} J_{ij} G_{jk} J_{lk} \frac{\partial B}{\partial y_l} = \frac{\partial A}{\partial y_i} G_{ik} \frac{\partial B}{\partial y_k} = \{A, B\}_{Q,P}. \quad (4.24)$$

The converse is also true since we may write the Jacobian as a 2 x 2 block matrix

$$J = \begin{pmatrix} J_1 & J_2 \\ J_3 & J_4 \end{pmatrix},$$

with

$$(J_1)_{ij} \equiv \frac{\partial Q_i}{\partial q_j}, \quad (J_2)_{ij} \equiv \frac{\partial Q_i}{\partial p_j}, \quad (J_3)_{ij} \equiv \frac{\partial P_i}{\partial q_j}, \quad (J_4)_{ij} \equiv \frac{\partial P_i}{\partial p_j},$$

i, j now running from 1 to n.
Then

$$(J G J^T) = \begin{pmatrix} \{Q_i, Q_j\}_{q,p} & \{Q_i, P_j\}_{q,p} \\ \{P_i, Q_j\}_{q,p} & \{P_i, P_j\}_{q,p} \end{pmatrix}, \quad (4.25)$$

so that, if the Poisson brackets are preserved, $\{A, B\}_{q,p} = \{A, B\}_{Q,P}$, the right hand side of eq. (4.25) is G, i.e. the transformation is canonical.

As a consequence of the above equivalence, one may adopt the following definition:

a transformation of the canonical variables is canonical if it leaves the Poisson brackets invariant, i.e. if, for given $q, p \to Q(q, p, t), P(q, p, t)$

$$\{Q_i, Q_j\}_{q,p} = 0, \quad \{P_i, P_j\}_{q,p} = 0, \quad \{Q_i, P_j\}_{q,p} = \delta_{ij}. \quad (4.26)$$

The above invariance property of the Poisson brackets may be exploited for simplifying the calculation of Poisson brackets.

Example 4.5. Using eq. (4.19), it is easy to check that $\{\mathbf{L}, F(r)\} = 0$, $r = \sqrt{\mathbf{x}^2}$.

The rather lengthy calculation in terms of Cartesian canonical variables may be replaced by a trivial one, by using canonical spherical variables for computing the Poisson bracket. In fact, in this case the only non-vanishing contribution to the Poisson brackets of $F(r)$ may come to $\partial F(r)/\partial r$ and, on the other side, from eq. (4.19) one has $\partial \mathbf{L}/\partial p_r = 0$.

Example 4.6. It easy to check that, for one particle, $\{L_i, T\} = 0$.

The computation is trivial by using Cartesian canonical variables; by eqs. (4.8), (4.12), (4.13) one immediately gets

$$\{L_i, \sum_l p_l p_l\} = \sum_{lk} \varepsilon_{ilk}(p_k p_l + p_l p_k) = 0.$$

Example 4.7. Consider the motion of a particle in a central potential $V(r)$ in a rotating frame $\mathcal{R}'$ with constant angular velocity $\boldsymbol{\omega}$ (as in Example 3.7). Then $\mathbf{L} \cdot \boldsymbol{\omega}$ is a constant of motion.

In fact, by using spherical variables with the z-axis in the direction of the rotation axis, one has $\mathbf{L} \cdot \boldsymbol{\omega} = p_\varphi \omega$, $H = T + V(r) - p_\varphi \omega$, and (by using the above result $\{L_i, T\} = 0\}$)

$$\{p_\varphi, H\} = \{p_\varphi, T + V - p_\varphi \omega\} = 0.$$

Example 4.8. Consider the case of Example 4.7 and a vector physical quantity $\mathbf{J}(\mathbf{x}, \mathbf{p})$. Determine its time variation in the reference frame $\mathcal{R}'$.

In $\mathcal{R}'$ the quantity $\mathbf{J}$ is described by the following function of the coordinates of the rotating frame $\mathbf{J}'(\mathbf{x}', \mathbf{p}') \equiv \mathbf{J}(\mathbf{x}(\mathbf{x}', t), \mathbf{p}(\mathbf{x}', \mathbf{p}', t))$ and therefore

$$\frac{d\mathbf{J}'}{dt} = \sum_i \left(\frac{\partial \mathbf{J}'}{\partial x_i'} \dot{x}_i' + \frac{\partial \mathbf{J}'}{\partial p_i'} \dot{p}_i' \right) = \{\mathbf{J}', H'\}_{x', p'}.$$

By the invariance of the Poisson brackets under canonical transformations, one may compute the above Poison brackets by using the canonical variables q, p and use the fact that $H' = H - \boldsymbol{\omega} \cdot \mathbf{L}$ (see Example 3.7). Then

$$\{\mathbf{J}', H'\}_{x',p'} = \{\mathbf{J}, H - \boldsymbol{\omega} \cdot \mathbf{L}\}_{x,p} = dJ/dt + \boldsymbol{\omega} \cdot \{\mathbf{L}, \mathbf{J}\}_{x,p}.$$

An interesting special case is provided by $\mathbf{J} = \mathbf{L}$; then in a rotating frame the time derivative of $\mathbf{L}$ is given by

$$\frac{d\mathbf{L}'}{dt} = \{\mathbf{L}, H\} - \boldsymbol{\omega} \wedge \mathbf{L}. \tag{4.27}$$

Example 4.9. Bloch equations. The Hamiltonian of a free symmetrical spinning top, in an inertial reference frame $\mathcal{R}$, is given by

$$H = T = \mathbf{J}^2/2I, \tag{4.28}$$

where I denotes its moment of inertia and $\mathbf{J}$ its angular momentum.

If the top carries a charge, a magnetic moment $\mathbf{M} = \mu\mathbf{J}$ is associated to its angular momentum, so that in the presence of a uniform magnetic field $\mathbf{B}$ the Hamiltonian gets an additional term $-\mathbf{M} \cdot \mathbf{B}$. Discuss the time evolution of $\mathbf{J}$.

By using the Poisson brackets one easily gets the time evolution in

$$\frac{dJ_i}{dt} = \mu\mathbf{B} \cdot \{\mathbf{J}, J_i\} = -\mu(\mathbf{B} \wedge \mathbf{J})_i.$$

Hence, in the reference frame $\mathcal{R}'$ rotating with angular velocity $\boldsymbol{\omega} = -\mu\mathbf{B}$, with respect to $\mathcal{R}$, by eq. (4.27), one has $dJ'_i/dt = 0$, i.e., in the reference frame $\mathcal{R}$, $\mathbf{J}$ undergoes a uniform precession, with angular velocity $\boldsymbol{\omega}$ around the axis in the direction of the magnetic field.

Remark 4.1. A similar structure arises also in the case in which one has a physical quantity $\mathbf{J}$, with components J_i, $i = 1, 2, 3$, to which are associated Poisson brackets of the same form as those of the angular momentum, eqs. (4.14), $\{J_i, J_j\} = \epsilon_{ijk}J_k$, without requiring the existence of corresponding canonical variables q, p.

This is the case of the intrinsic angular momentum or *spin* **S** of a particle, the only properties which enter being the Poisson brackets

$$\{\mathcal{S}_i, \mathcal{S}_j\} = \varepsilon_{ijk}\,\mathcal{S}_k, \tag{4.29}$$

characteristic of an angular momentum.

If a magnetic moment $\mathbf{M} = \mu\mathbf{S}$ is associated to **S**, in the presence of a magnetic field one has a coupling with a magnetic field given by $\mu\,\mathbf{S}\cdot\mathbf{B}$ and the time evolution is given by equations of the same form of the Bloch equations and one has a precession as above.

This further shows the power of the Hamiltonian mechanics over the Newtonian Cartesian mechanics, covering cases which do not have an analog in terms of Cartesian coordinates.

Remark 4.2. It is important to stress that the canonical structure, and in particular the canonicity of a transformation, is *independent of the dynamics*, i.e. of the possible choice of the Hamiltonian. It characterizes the geometrical structure of the manifold with local coordinates q, p.

5

Generation of canonical transformations

5.1 Alternative characterization of canonical transformations

The characterization of canonical transformations discussed in the previous Chapter, in terms of eqs. (3.15) or of the invariance of the Poisson brackets, is on one side rather simple, but on the other side is not constructive.

In the following, we shall discuss a constructive strategy by characterizing (generating) functions, with the property that, through simple operations involving their derivatives, they define transformations of the canonical variables, which are guaranteed of being canonical.

To this purpose it is convenient to have the following alternative characterization of canonical transformations.

A transformation of the canonical variables

$$q, p \to Q(q, p, t), \quad P(q, p, t)$$

is equivalently identified by the inverse formulas

$$q_i = q_i(Q, P, t), \quad p_i = p_i(Q, P, t). \tag{5.1}$$

and more generally by assigning $2n$ invertible relations between the new and the old canonical variables.

© Springer International Publishing AG 2018
F. Strocchi, *A Primer of Analytical Mechanics*, UNITEXT for Physics,
https://doi.org/10.1007/978-3-319-73761-4_5

For example, if the $2n$ coordinates q, Q are functionally independent one may express the other coordinates in terms of them and the canonical transformation (5.1) may be equivalently characterized by the following $2n$ (invertible relations)

I)

$$p_i = p_i(q, Q, t), \quad P_i = P_i(q, Q, t), \quad i = 1, ...n. \tag{5.2}$$

In fact, by inverting the second n relations one may obtain the q's as functions of Q, P: $q_i = q_i(Q, P, t)$ and substituting such expressions in the first n relations one obtains also the p_i's as functions of Q, P. In this way one recovers the form of eq. (5.1).

In a similar way, according as the $2n$ coordinates (q, P), or (p, Q), or (p, P), are functionally independent one may correspondingly characterize the canonical transformation of eq. (5.1) by giving one of the following set of (invertible) relations

II)

$$p_i = p_i(q, P, t), \quad Q_i = Q_i(q, P, t), \quad i = 1, ...n; \tag{5.3}$$

III)

$$q_i = q_i(p, Q, t), \quad P_i = P_i(p, Q, t), \quad i = 1, ...n; \tag{5.4}$$

IV)

$$q_i = q_i(p, P, t), \quad Q_i = Q_i(p, P, t), \quad i = 1, ...n. \tag{5.5}$$

The crucial next step is to formulate the canonicity condition in terms of the relations I)-IV).

We start by case **I**. Then, as we shall argue below, the transformation defined by eq. (5.2) is canonical if and only if there exists a function $F_I = F_I(q, Q, t)$ such that

$$p_i(q, Q, t) = \frac{F_I(q, Q, t)}{\partial q_i}, \quad P_i(q, Q, t) = -\frac{F_I(q, Q, t)}{\partial Q_i}, \quad i = 1, ...n. \tag{5.6}$$

Thus, from a constructive point of view, any regular function of $2n$ independent variables q, Q, defines a canonical transformation, through eqs. (5.6).

The function F_I is called the *generating function of the canonical transformation*.

The function F_I also allows to write the Hamiltonian $H_{Q,P}$ for the new variables:

$$H_{Q,P}(Q, P, t) = H(q(Q, P, t), p(Q, P, t), t) + \partial F_I/\partial t. \tag{5.7}$$

The canonicity conditions for the relations (5.3)-(5.5) require the existence of functions $F_{II}(q, P, t), F_{III}(p, Q, t), F_{IV}(p, P, t)$ such that

II)

$$p_i(q, P, t) = \frac{\partial F_{II}}{\partial q_i}, \quad Q_i(q, P, t) = \frac{F_{II}}{\partial P_i}; \tag{5.8}$$

III)

$$q_i(p, Q, t) = -\frac{\partial F_{III}}{\partial p_i}, \quad P_i(p, Q, t) = -\frac{F_{III}}{\partial Q_i}; \tag{5.9}$$

IV)

$$q_i(p, P, t) = -\frac{\partial F_{IV}}{\partial p_i}, \quad Q_i(p, P, t) = \frac{\partial F_{IV}}{\partial P_i}. \tag{5.10}$$

We shall prove that any of the above set of equations, eqs. (5.6) or (5.8) or (5.9) or (5.10), imply that the corresponding transformation is canonical.

As discussed in Section 4.4, it is enough to show that the Poisson brackets are invariant.

To this purpose we start by considering the case of eqs. (5.6); one has

$$\frac{\partial P_j}{\partial p_k}|_q = \frac{\partial P_j}{\partial Q_l}|_q \frac{\partial Q_l}{\partial p_k}|_q; \quad \frac{\partial P_j}{\partial q_k}|_p = \frac{\partial P_j}{\partial q_k}|_Q + \frac{\partial P_j}{\partial Q_l}|_q \frac{\partial Q_l}{\partial q_k}|_p.$$

where the sum over repeated indices is understood and $|_q, |_p$ denote that in the derivation the variable q, respectively p, has to be kept fixed. Then, by using the above equations one gets

$$\{Q_i, P_j\} = \frac{Q_i}{\partial q_k}|_p \frac{\partial P_j}{\partial p_k}|_q - \frac{Q_i}{\partial p_k}|_q \frac{\partial P_j}{\partial q_k}|_p = -\frac{\partial Q_i}{\partial p_k}|_q \frac{\partial P_j}{\partial q_k}|_Q =$$

$$\frac{\partial Q_i}{\partial p_k}|_q \frac{\partial}{\partial q_k}|_Q \frac{\partial F_I}{\partial Q_j}|_q = \frac{\partial Q_i}{\partial p_k}|_q \frac{\partial p_k}{\partial Q_j}|_q = \delta_{ij}.$$

In a similar way one recovers the other canonical Poisson brackets.

The other cases **II-IV** may be treated by an analogous pattern. We skip the proof of the converse, namely that if the transformation is canonical there are functions $F_I, F_{II}, F_{III}, F_{IV}$ such that eqs. (5.6), (5.8), (5.9), (5.10) hold.

In all cases the Hamiltonian for the new canonical variables $H_{Q,P}$ is obtained from $H_{q,p}$ as in eq. (5.7), with F_I replaced by the corresponding generating functions.

If the relations between the old and the new canonical variables are not explicitly dependent on time, then so are also the generating functions F and therefore the Hamiltonian (which may explicitly depend on time) transforms covariantly under the transformation, as the Lagrangian function does for a change of Lagrangian variables.

The *Hamiltonian is invariant under a canonical transformation* if $H_{Q,P}$ and $H_{q,p}$ are the same function of their arguments.

*Remark 5.1 It is worthwhile to remark that the characterization of the canonical transformations in terms of generating functions has the advantage of providing the transformation of the Hamiltonian, an information not directly given by the invariance of the Poisson brackets.

If the relation between the old and the new canonical variables is explicitly time dependent the Hamiltonian is not covariant under the transformation. In fact, by eqs. (4.12), putting $K \equiv H_{Q,P}$, one has

$$\dot{Q}_i = \{Q_i, K\}_{Q,P} = \partial K/\partial P_i, \quad \dot{P}_i = \{P_i, K\}_{Q,P} = -\partial K/\partial Q_i \quad (5.11)$$

On the other hand, as functions of the old variables Q_i, P_i satisfy

$$\dot{Q}_i = \{Q_i, H\}_{q,p} + \partial Q_i/\partial t, \quad \dot{P}_I = \{P_i, H\}_{q,p} + \partial P_i/\partial t \quad (5.12)$$

and, by the invariance of the Poisson brackets (eqs. (4.21), (4.22)),

$$\{Q_i, H\}_{q,p} = \{Q_i, H\}_{Q,P} = \partial H/\partial P_i,$$

$$\{P_i, H\}_{q,p} = \{P_i, H\}_{Q,P} = -\partial H/\partial Q_i.$$

Then, by comparing eqs. (5.11), (5.12), one has

$$\frac{\partial Q_i}{\partial t} = \frac{\partial (K - H)}{\partial P_i}, \quad \frac{\partial P_i}{\partial t} = -\frac{\partial (K - H)}{\partial Q_i} \quad (5.13)$$

and $K - H \neq 0$ if there is an explicit time dependence in the expression of the new canonical variables in terms of the old ones.

Moreover, the left hand sides of eqs. (5.13) depend on the functions $Q(q, p, t)$, $P(q, p, t)$, which define the transformation, but not on their dynamics, defined by the Hamiltonians H, K. Therefore, the solution $G \equiv K - H$ of eqs. (5.13) is independent of the dynamics, being completely determined by the transformation $q, p \to Q(q, p, t)$, $P(q, p, t)$.

Now, a transformation $q, p \to Q, P$, with q, p treated as independent variables defines a transformation $q, \dot{q} \to Q, \dot{Q}$ and a corresponding Lagrangian $L'(Q, \dot{Q}, t)$ such that $K = \sum_i P_i \dot{Q}_i - L'(Q, \dot{Q}, t)$. Hence, one has

$$K - H = \sum_i P_i \dot{Q}_i - L' - \sum_i p_i \dot{q}_i + L.$$

Since L' and L must give the same Euler equations of motion in terms of the variables $q, \dot{q}$, by the argument of Section 2.9 they must differ by a total derivative $d\mathcal{F}/dt$. Then, if q_i, Q_i are functionally independent one may express $\mathcal{F}$ as a function of q, Q and

$$K - H = \sum_i (P_i \dot{Q}_i - p_i \dot{q}_i) + L - L' =$$

$$= \sum_i \left(P_i \dot{Q}_i - p_i \dot{q}_i + \frac{\partial \mathcal{F}}{\partial q_i} \dot{q}_i + \frac{\partial \mathcal{F}}{\partial Q_i} \dot{Q}_i \right) + \frac{\partial \mathcal{F}}{\partial t}.$$

By the above remark, $G \equiv K - H$ does not depend on $\dot{q}_i, \dot{Q}_i$, which involve the dynamics of these two variables, and therefore one must have

$$P_i(q, Q, t) = -\frac{\partial \mathcal{F}(q, Q, t)}{\partial Q_i}, \quad p_i(q, Q, t) = \frac{\partial \mathcal{F}(q, Q, t)}{\partial Q_i}, \tag{5.14}$$

$$K - H = \frac{\partial \mathcal{F}(q, Q, t)}{\partial t}. \tag{5.15}$$

A comparison with eqs. (5.6) shows that $\mathcal{F}(q, Q, t)$ may be identified with the generating function $F_I(q, Q, t)$ and eq. (5.7) follows.

A similar argument may be used by choosing q, P or p, Q or p, P as independent variables leading to the generating functions F_{II}, F_{III}, F_{IV}; in this way one proves the corresponding equations for the transformation of the Hamiltonian.

Example 5.1 Determine the canonical transformations defined by the following generating functions:

$$1) \quad F_1 = \sum_i q_i\, Q_i, \qquad 2) \quad F_2 = \sum_i q_i\, P_i,$$

$$3) \quad F_3 = -\sum_i p_i\, Q_i, \qquad 4) \quad F_4 = -\sum_i q_i\, P_i.$$

1). For F_1 one easily gets

$$p_i = \frac{\partial F_1(q, Q)}{\partial q_i} = Q_i, \quad P_i = -\frac{\partial F_1(q, Q)}{\partial Q_i} = -q_i,$$

i.e. the transformation exchanges the "position" variables q with the momenta p.

2) For F_2 one has

$$p_i = \frac{\partial F_2(q, P)}{\partial q_i} = P_i, \quad Q_i = \frac{\partial F_2(q, P)}{\partial P_i} = q_i,$$

i.e. the F_2 defines the *identity transformation*.

3) The same conclusion is easily obtained for the canonical transformation defined by F_3.

4) For F_4 the transformation amounts to changing the sign of both the q's and the p's. In the case of Cartesian coordinates such a transformation has the meaning of the *parity* transformation.

The above canonical transformations further show that in the Hamiltonian formulation the canonical variables are *all* on the same footing, the separation of positions and momenta being not stable under canonical transformations.

Thus, whereas in the Lagrangian formulation the basic space is the n-dimensional space of the positions (all on the same footing), in the Hamiltonian formulation the basic space if the so-called **phase space** of the $2\,n$ canonical variables q, p.

Example 5.2. Determine the generating functions of the following canonical transformations, for simplicity for one-particle coordinates

a) *Hamiltonian Galilei transformation* of the Hamiltonian variables

$$x_i' = x_i - w_i t, \quad p_i' = p_i - m\, w_i, \quad i = 1, 2, 3; \qquad (5.16)$$

b) *rotating frame transformation*, corresponding to Lagrangian coordinates in the frame $\mathcal{R}'$ which rotates with uniform angular velocity ω (see Example 3.7)

$$\rho' = \rho, \quad \varphi' = \varphi - \omega t, \quad z' = z; \quad p_{\rho'} = p_\rho, \quad p_{z'} = p_z, \quad p_{\varphi'} = p_\varphi. \qquad (5.17)$$

Case a). For simplicity we discuss the case of one single degree of freedom, with $m = 1$.

Both the pairs (x, x') and the pair (p, p') do not consist of functionally independent variables, whereas so do the pairs (x, p') and (p, x'); therefore II) and III) apply.

The generating function F_{II} may be determined by solving eqs. (5.8)

$$\frac{\partial F_{II}(x, p')}{\partial x} = p = p' + v, \quad \Rightarrow \quad F_{II} = (p' + v)x + f(p')$$

$$\frac{\partial F_{II}}{\partial p'} = x' = x + v t, \quad \Rightarrow \quad f(p') = -p'\, v t.$$

According to eq. (5.7), $H'(x', p') = H(x(x'), p(p')) - p'v$ and in the free case $H'(x', p') = \frac{1}{2}(p' + v)^2 - p'\, v$.

The generating function $F_{III}(p, x')$ is obtained by solving eqs. (5.9): the first equation gives $F_{III} = -p(x' + v t) + g(x')$ and the second implies $g(x') = x'v$.

Case b). In this case, the pairs of functionally independent coordinates are the pair (q, P) and the pair (p, Q). It easy to prove that the generating function corresponding to case II is

$$F_{II} = \rho\, p_{\rho'} + z\, p_{z'} + \varphi\, p_{\varphi'} - \omega t\, p_{\varphi'} = \rho\, p_\rho + z\, p_z + \varphi\, p_\varphi - \omega t\, p_\varphi,$$

and $H' = H - \omega\, \varphi = H - \boldsymbol{\omega} \cdot \mathbf{L}$.

The generating function F_{III} may be determined in a similar way:

$$F_{III} = -(p_\rho\, \rho + p_z\, z + p_\varphi\, (\varphi' + \omega t)).$$

Remark 5.2. It is worthwhile to note that there is a certain ambiguity in identifying the transformation of the canonical variables which correspond to changes to moving frame coordinates.

In fact, according to Section 3.3, e.g. for a free particle of unit mass, the *Galilei transformation of Lagrangian coordinates* $x_i' = x_i - w_i t$, $\dot{x}_i = \dot{x}_i - w_i$ leads to $p_i' \equiv \partial L/\partial \dot{x}_i' = p_i = \dot{x}_i$.

On the other hand, the transformation to a moving frame corresponds to a *Hamiltonian Galilei transformation* and one may represent it as $Q_i = x_i - w_i t$, $P_i = p_i - w_i$.

The apparent conflict is resolved by noting that the canonical momentum is not uniquely identified, since it changes if a total time derivative is added to the Lagrangian (which does not change the equation of motion). Both transformations are canonical, since they leave the Poisson brackets invariant, but the physical interpretation of the canonical momenta is very different.

In the first case, the Hamiltonian in the new canonical variables is $H_{x'p'} = [\frac{1}{2}p_i'^2 - p_i' v_i]$ (see Example 3.6 or use the generator F_{II} and the equation corresponding to eq. (5.7)), so that $\dot{x}_i' = p_i' - w_i$ and p_i' is not the velocity in the moving frame.

In the second case, the Hamiltonian is $H_{Q,P} = [\frac{1}{2}(P_i + w_i)^2 - P_i w_i]$ and $\dot{Q}_i = P_i$, i.e. the canonical variables Q_i, P_i have the physical interpretation of position and momentum in the moving frame.

Example 5.3. Determine the generating function of the canonical transformation (3.10), (3.11)

$$Q_i = Q_i(q, t), \qquad P_i = \sum_j p_j \, \partial q_j/\partial Q_i. \tag{5.18}$$

The relation between the p's and the P's may also be written as $p_i = \sum_j P_j \, \partial Q_j/\partial q_i$; then, by using q, P as independent variables (case II), from the first of eqs. (5.8) one gets

$$F_{II} = \sum_j Q_j(q, t) \, P_j + g(P, t)$$

and the second gives $g(P, t) = const.$

5.2 Extended canonical transformations

The characterization of transformations $(q, p) \rightarrow (Q, P)$ by the require-
ment that there exists a Hamiltonian function $H_{Q,P}$, such that the
Hamilton equations hold for the new canonical variables, leaves open
the possibility of a scale factor λ in the relation between the two Hamiltonians.

Equivalently, one may correspondingly consider transformations of
the canonical variables which leave the Poisson brackets invariant apart
from an overall scale factor (**extended canonical transformations**):

$$\{Q_i, P_k\}_{q,p} = \lambda\, \delta_{ik}, \quad \{Q_i, Q_k\}_{q,p} = \{P_i, P_k\}_{q,p} = 0. \quad (5.19)$$

It is worthwhile to see how this more general case reflects in the corresponding generating functions. One has the following corresponding
cases:

I.
$$p_i(q, Q, t) = \frac{\partial F_I(q, Q, t)}{\partial q_i}, \quad P_i = -\lambda \frac{\partial F_I(q, Q, t)}{\partial Q_i}, \quad (5.20)$$
$$H_{Q,P} = \lambda\, H(q(Q, P, t), p(Q, P, t)) + \lambda\, \partial F_I / \partial t,$$

II).
$$p_i(q, P, t) = \frac{\partial F_{II}}{\partial q_i}, \quad Q_i(q, P, t) = \lambda \frac{F_{II}}{\partial P_i}, \quad (5.21)$$
$$H_{Q,P} = \lambda\, H(q, p, t) + \lambda\, \partial F_{II} / \partial t;$$

III).
$$q_i(p, Q, t) = -\frac{\partial F_{III}}{\partial p_i}, \quad P_i(p, Q, t) = -\lambda \frac{F_{III}}{\partial Q_i}, \quad (5.22)$$
$$H_{Q,P} = \lambda\, H(q, p, t) + \lambda\, \partial F_{III} / \partial t;$$

IV).
$$q_i(p, P, t) = -\frac{\partial F_{IV}}{\partial p_i}, \quad Q_i(p, P, t) = \lambda \frac{\partial F_{IV}}{\partial P_i}, \quad (5.23)$$
$$H_{Q,P} = \lambda\, H(q, p, t) + \lambda\, \partial F_{IV} / \partial t.$$

Example 5.4. The transformation

$$Q_i = q_i, \qquad P_i = -p_i, \tag{5.24}$$

is not a (standard) canonical transformation, but it is an extended canonical transformation.

In fact, one has

$$\{Q_i, P_j\}_{q,p} = -\{q_i, p_j\}_{q,p} = -\delta_{ij}.$$

This corresponds to $\lambda = -1$ in eq. (5.19).
The corresponding transformation for Cartesian variables reads:

$$x_i' = x_i, \qquad \dot{x}_i' = -\dot{x}_i$$

and has the physical interpretation of *time reversal*.
The Lagrangian functions which are quadratic functions of the $\dot{x}_i$ are clearly invariant under such a transformation and therefore the time evolution of the corresponding systems is *reversible*.

Example 5.5. The transformation

$$Q_i = s\, q_i, \qquad P_i = s\, p_i, \qquad s \in \mathbf{R}, \tag{5.25}$$

is an extended canonical transformation with $\lambda = s^2$.

It follows immediately from the check of the Poisson brackets, according to eq. (5.19).
For Cartesian coordinates the transformation has the physical meaning of a *scale transformation*, corresponding to a change of the unit of measurements of the space coordinates.

Example 5.5' More generally, the transformation

$$Q_i = \Lambda_{ij}\, q_j, \qquad P_i = \Sigma_{ij} p_j,$$

with Λ a real, symmetric, strictly positive matrix, is an extended canonical transformation if $\Lambda \Sigma^T = \lambda \mathbf{1}$, $\lambda \in \mathbf{R}$ and a standard canonical transformation if $\lambda = 1$.

5.3 Generators of continuous groups of canonical transformations

Clearly, the canonical transformations form a group and it is instructive to analyze the continuous (Lie) subgroups connected with the identity.

The corresponding generating functions may be written in the form II) or III), since then their being connected with the identity may be described by an additional (continuous) dependence on the group parameters α, with the property that for $\alpha = 0$ one obtains the identity transformation:

$$F_{II}(q, P, \alpha), \quad F_{II}(q, P, \alpha = 0) = \sum_i q_i \, P_i;$$

$$F_{III}(p, Q, \alpha), \quad F_{III}(p, Q, \alpha = 0) = -\sum_i p_i \, Q_i.$$

We further assume that the above generating functions admit a power series expansion in a small neighborhood $\mathcal{N}$ of $\alpha = 0$.
E. g., for a one-parameter group one has

$$F_{II}(q, P, \alpha) \simeq \sum_i q_i \, P_i + \alpha \, G(q, P), \quad \alpha \in \mathcal{N}. \tag{5.26}$$

Hence, the corresponding transformation is

$$Q_i = \frac{F_{II}}{\partial P_i} = q_i + \alpha \frac{\partial G(q, P)}{\partial P_i},$$

$$p_i = \frac{\partial F_{II}}{\partial q_i} = P_i + \alpha \frac{\partial G(q, P)}{\partial q_i} \quad \Rightarrow \quad P_i = p_i - \alpha \frac{\partial G}{\partial q_i}. \tag{5.27}$$

The above considerations may be extended to the continuous groups of extended canonical transformations connected with the identity.

Example 5.6. The generating functions of the following continuous transformations
1) the *translations* of the i-th coordinate:

$$Q_i = q_i + \alpha, \quad Q_j = q_j, \; j \neq i, \quad P_k = p_k, \; \forall k \tag{5.28}$$

2) the infinitesimal euclidean *rotations* around the x_3- axis,

$$x_1' = x_1 - \alpha x_2, \quad x_2' = x_2 + \alpha x_1, \quad x_3' = x_3,$$

$$P_x = p_x - \alpha p_y, \quad P_2 = p_2 + \alpha p_1, \quad P_3 = p_3, \tag{5.29}$$

may be written in the form (5.26).

In fact, the corresponding generating functions are, respectively,

$$F_{II}(q, P) = \sum_j q_j\, P_j + \alpha P_i = \sum_j q_j\, P_j + \alpha p_i,$$

$$F_{II}(\mathbf{x}, \mathbf{P}) = (x_1 - \alpha x_2)P_1 + (x_2 + \alpha x_1)\, P_2 + x_3\, P_3 =$$
$$= \sum_i x_i P_i + \alpha(x_1\, p_2 - x_2\, p_1),$$

Example 5.7. The *scale transformations* considered in Example 5.5, eq. (5.25), with $s = e^\alpha$, $\alpha \in \mathbf{R}$, define a (one-parameter) continuous group of extended canonical transformations, with group parameter α, the identity corresponding to $\alpha = 0$.

In fact, according to eqs. (5.26) the generating function $F_{II}(q, P) = \sum_i e^{-\alpha} q_i\, P_i$ generates the transformation

$$Q_i = e^\alpha\, q_i, \qquad P_i = e^\alpha\, p_i, \quad \alpha \in \mathbf{R},$$

in agreement with eqs. (5.19), with $\lambda = e^{2\alpha}$. Moreover, for α in a neighborhood of $\alpha = 0$, the generating function it is of the form of eq. (5.26). Clearly, a similar construction of the generating functions may easily be done for the (one-parameter) continuous group of standard canonical transformations

$$Q_i = e^\alpha\, q_i, \qquad P_i = e^{-\alpha}\, p_i, \quad \alpha \in \mathbf{R};$$

in fact, the corresponding generating function is of the form of eq. (5.26)

$$F_{II}(q, P, \alpha) = e^\alpha \sum_i q_i\, P_i, \qquad F_{II}(q, P, \alpha) \simeq \sum_i q_i\, P_i + \alpha \sum_i q_i\, P_i.$$

Given a continuous group of canonical transformations, for each one-parameter subgroup, the *infinitesimal variations of the canonical variables* is obtained by expanding the corresponding generating function in the neighborhood of the identity of the group parameter, as implicitly given by eqs. (5.27).

In fact, the group structure implies that the equations for p_i and Q_i may be inverted to first order in $\alpha = \varepsilon \ll 1$ and one has

$$p_i \simeq P_i + \varepsilon \frac{\partial G(q, P)}{\partial q_i} = P_i + \varepsilon \frac{\partial G(q, p)}{\partial q_i} + O(\varepsilon^2), \qquad (5.30)$$

$$Q_i \simeq q_i + \varepsilon \frac{\partial G(q, P)}{\partial P_i} = Q_i + \varepsilon \frac{\partial G(q, p)}{\partial p_i} + O(\varepsilon^2), \qquad (5.31)$$

where we have expanded $\partial G(q, P)/\partial q_i$ and $\sum_j (\partial G(q, P)/\partial p_j) \partial p_j/\partial P_i$ in the neighborhood of $P_k = p_k$, $\forall k$, keeping only the first orders in ε.

The function $G(q, p)$ which appears in eqs. (5.30), (5.31), is called the **generator of the infinitesimal transformation**.

The above somewhat laborious derivation may be greatly simplified by using the Poisson brackets; actually in this way one may easily obtain the infinitesimal variation of any physical quantity described by a (regular) function of the canonical variables, $A(q, p)$.

To this purpose we note that (to first order in ε) eqs. (5.30), (5.31) may be written in the following form

$$\delta p_i = P_i - p_i = \varepsilon \{p_i, G\}_{q,p}; \quad \delta q_i = Q_i - q_i = \varepsilon \{q_i, G\}_{q,p}, \qquad (5.32)$$

i.e. the infinitesimal variations of q, p are respectively given by their Poisson brackets with the generator $G(q, p)$.

Actually, any (regular) function $G(q, p)$ generates an infinitesimal canonical transformation through eqs. (5.32).

To this purpose, we have to check that, up second order in ε, the Poisson brackets of the variables

$$Q_i \equiv q_i + \varepsilon \{q_i, G\}_{q,p} \quad P_i \equiv p_i + \varepsilon \{p_i, G\}_{q,p}$$

are canonical.

In fact, by using eqs. (4.12), to first order in ε one has

$$\{Q_i, Q_j\} = \{q_i + \varepsilon \{q_i, G\}, q_j + \varepsilon \{q_j, G\}\} = O(\varepsilon^2),$$

$$\{P_i, P_j\} = \{p_i + \varepsilon \{p_i, G\}, p_j + \varepsilon \{p_j, G\}\} = O(\varepsilon^2),$$

$$\{Q_i, P_j\} = \{q_i + \varepsilon \{q_i, G\}, p_j + \varepsilon\{p_i, G\}\} =$$

$$= \{q_i, p_j\} + \varepsilon(\{\{q_i, G\}, p_j\} + \{q_i, \{p_j, G\}\}) + O(\varepsilon^2) = \delta_{ij} + O(\varepsilon^2).$$

The important role of the generator of a canonical transformation, $G(q, p)$, will be further displayed below and it is useful to note that its identification does not necessarily requires the determination of the generating function and of its first order expansion in the group parameter, being possible to directly obtaining it from the infinitesimal transformation of the canonical variables; in fact, it may be identified as the function of the canonical variables q, p, whose Poisson brackets give δq_i, δp_i, according to eqs. (5.32).

In fact, a function of the canonical variables $A(q, p)$ is uniquely determined by its Poisson brackets with q, p, apart from a constant: if $A(q, p)$, $B(q, p)$ have the same Poisson brackets with the canonical variables. i.e.

$$\{A(q, p), q_i\} = F_i(q, p) = \{B(q, p), q_i\},$$

$$\{A(q, p), p_i\} = G_i(q, p) = \{B(q, p), p_i\}, \quad \forall i,$$

then, by eqs. (4.12), $A(q, p) - B(q, p)$ is independent of p_i and of q_i, and therefore is a constant.

*__Example 5.8__ By using the condition of invariance of the Poisson brackets, rather than eqs. (5.30), (5.31), show that the infinitesimal canonical transformations are of the form (5.32).

$$\delta q_i = \varepsilon \frac{\partial F(q, p)}{\partial p_i}, \quad \delta p_i = -\varepsilon \frac{\partial F(q, p)}{\partial q_i},$$

with a suitable $F(q, p)$, which is therefore the generator of the transformations.

In fact, quite generally, infinitesimal transformations are of the form

$$q_i' = q_i + \varepsilon F_i(q, p), \quad p_i' = p_i + \varepsilon E_i(q, p).$$

The canonicity condition $\{q'_i,\, q'_j\} = 0$ gives

$$\frac{\partial F_j}{\partial p_i} - \frac{\partial F_i}{\partial p_j} = 0,$$

and by Poincaré Lemma, there exists a function F, such that (locally) $F_i = \partial F / \partial p_i$. Similarly, the condition $\{p'_i,\, p'_j\} = 0$ gives $E_i = \partial E / \partial q_i$. Finally, the condition $\{q'_i,\, p'_j\} = \delta_{ij}$ gives

$$\frac{\partial^2 (F + E)}{\partial q_j\, \partial p_i} = 0,$$

which implies $F + E = f(q) + g(p)$ and by a redefinition $F \to F' = F - f$, $E \to E' = E - g$, which does not change the equations for q'_i, p'_i, in terms of q_i, p_i, one obtains $E' = -F'$

In terms of the Poisson brackets one may also determine the infinitesimal variation of the generic function $A(q, p)$.

In fact, one has (with the sum over repeated indices understood)

$$\delta A = \frac{\partial A}{\partial q_i}\, \delta q_i + \frac{\partial A}{\partial p_i}\, \delta p_i =$$

$$\varepsilon \frac{\partial A}{\partial q_i} \{q_i,\, G\} + \varepsilon \frac{\partial A}{\partial p_i} \{p_i,\, G\} = \varepsilon\{A,\, G\}, \qquad (5.33)$$

so that *the Poisson bracket with the generator $G(q, p)$ give the infinitesimal variation* of a (regular) function $A(q, p)$.

In particular, A is invariant under an infinitesimal transformation if its Poisson bracket with the corresponding generator vanishes. The (Lie) group structure implies that in this case A is invariant under the full (one parameter Lie) group of transformations, since any finite transformation may be obtained by iterating infinitesimal transformations.

A very relevant physical consequence it that the invariance of the Hamiltonian under a (one-parameter) continuous group of canonical transformations implies the vanishing of its infinitesimal variation and by eq. (5.33) of its Poisson bracket $\{H,\, G\} = 0$ with the generator G. Hence, if $G = G(q, p)$, with no explicit time dependence, G is a constant of motion, (**strict relation between symmetries of the Hamiltonian and constants of motion**).

Example 5.9. It is instructive to determine the generators of the following important canonical transformations of the Cartesian coordinates of a system of N particles.

a) *Space translations*

We denote by $\mathbf{x}^\alpha$ the position of the α-th particle in $\mathbf{R}^3$ and by x_i^α, $i = 1, 2, 3$, its Cartesian components; a space translation in the i-th direction is defined by

$$\delta x_i^\alpha = a, \quad \delta x_j^\alpha = 0, \quad j \neq i, \quad \delta p_k^\alpha = 0; \quad \forall \alpha.$$

According to Example 5.6, the generating function is

$$F_{II} = \sum_{\alpha,j} x_j^\alpha \, P_j^\alpha + a \sum_\alpha P_i^\alpha.$$

Then, according to eqs. (5.30), (5.31), the generator is the i-th component of the *total moment* $\sum_\alpha P_i^\alpha = \sum_\alpha p_i^\alpha$. This generator may also be directly identified as the function whose Poisson brackets with $\mathbf{x}^\alpha, \mathbf{p}^\alpha$ give $\delta \mathbf{x}^\alpha, \delta \mathbf{p}^\alpha$.

Then, if the Hamiltonian is invariant under all translations, the total momentum is a constant of motion

$$0 = \delta H = \{H, \mathbf{P}\} = -d\mathbf{P}/dt.$$

b) *Rotations in $\mathbf{R}^3$*

For each particle coordinates the infinitesimal rotations around the third axis are defined by eqs. (5.29), which identify the third component of the *(total) angular momentum* as the generator. This is also a direct consequence of the Poisson brackets (4.13)

$$\delta x_i = \alpha \, \varepsilon_{ijk} x_k = \alpha \{x_i, L_j\}.$$

The angular momentum $\mathbf{J}$ is also the generator of rotations for a free symmetrical top (see Example 4.9). This follows from the transformation of the angular momentum under infinitesimal rotations and eq. (4.14)

$$\delta J_i = \varepsilon \, \varepsilon_{ijk} J_k = \varepsilon \{J_i, J_j\}.$$

The invariance of the Hamiltonian under rotations implies that the (total) angular momentum is a constant of motion.

c) *Time translations*

Infinitesimal time translations, $\alpha = \Delta t \ll 1$, are defined by

$$\delta q_i = \Delta t\, \dot{q}_i = \Delta t\, \{q_i, H\}, \quad \delta p_i = \Delta t\, \dot{p}_i = \Delta t\, \{p_i, H\},$$

where eq. (4.4) has been used.

Thus, the *Hamiltonian* is the *generator of time translations*.

d) *Galilei transformations*

As discussed in Remark 5.2, one may consider the following transformation of the canonical variables as describing Galilei Transformations

$$x_1' = x_1 - w_1\, t, \quad x_2' = x_2 - w_2\, t, \quad x_3' = x_3 - w_3\, t, \quad \Leftrightarrow \quad \delta x_i = -w_i\, t$$

$$p_1' = p_1 - m\, w_1, \quad p_2' = p_2 - m\, w_2, , \quad p_3' = p_3 - m\, w_3, \quad \Leftrightarrow \quad \delta p_i = -m\, w_i,$$

since they may be interpreted as relating canonical variables in an inertial frame $\mathcal{R}$ to those in a frame $\mathcal{R}'$ moving with constant velocity $\mathbf{w}$ with respect to $\mathcal{R}$.

By using the Poisson brackets, it is easy to check that

$$\mathbf{G} = -\mathbf{p}\, t + m\, \mathbf{x}, \tag{5.34}$$

is the *generator of the above (Galilei) transformations*.

Example 5.10. The invariance of the Hamiltonian under rotations implies that the angular momentum is a constant of motion.

For concreteness we consider the Hamiltonian of one particle subject to a central potential $V = V(r)$.

The obvious invariance under rotations may be explicitly checked by computing the Poisson brackets with the generators L_i; in fact, by the computations of Examples 4.5, 4.6, one has

$$\{H, L_i\} = 0,$$

and L_i is a constant of motion.

5.4 Symmetries and conservation laws. Noether theorem

The deep discovery of Emmy Noether was that to each continuous symmetry of the dynamics there corresponds a conserved quantity. This relation between symmetries and conservations laws has become one of the cornerstone of modern theoretical physics.

5.4.1 Noether theorem: Lagrangian formulation

In the Lagrangian formulation, according to the discussion of Section 2.9 we are led to consider as *symmetries of the dynamics* those transformations of the Lagrangian coordinates $q \to Q$, $\dot{q} \to \dot{Q}$, such that the *Lagrangian is invariant up to a total derivative*:

$$L'(Q, \dot{Q}, t) \equiv L(q(Q), \dot{q}(Q, \dot{Q}), t) = L(Q, \dot{Q}, t) - \frac{d\,\mathsf{G}'(Q)}{d\,t}. \qquad (5.35)$$

Equivalently, one has $(G(q) \equiv \mathsf{G}'(Q(q)))$

$$L(Q(q, t), \dot{Q}(q, \dot{q}, t), t) - L(q, \dot{q}, t) = \frac{d\,\mathsf{G}(q)}{dt}. \qquad (5.36)$$

When one has a one-parameter continuous group of transformations the invariance of the Lagrangian up to a total derivative, eq.(5.36), may checked at the infinitesimal level (omitting the possible time dependence of F, G and using $\mathsf{G}(q, \varepsilon) = \varepsilon G(q) + O(\varepsilon^2)$):

$$\delta q_i = \varepsilon\, F_i(q, t), \quad \delta \dot{q}_i = \varepsilon\, \frac{d\,F_i(q, t)}{dt}, \qquad (5.37)$$

$$\delta L \equiv L(q + \delta q, \dot{q} + \delta \dot{q}, t) - L(q, \dot{q}, t) = \varepsilon\, \frac{dG(q)}{dt}. \qquad (5.38)$$

Theorem 5.4.1 (Noether theorem). *To each one-parameter group $\mathcal{G}$ of symmetries of the dynamics, which therefore, in the Lagrangian formulation, define infinitesimal transformations satisfying eqs. (5.37), (5.38), there corresponds a conserved quantity*

$$\mathcal{Q}(q, \dot{q}, t) \equiv \sum_i \frac{\partial L}{\partial \dot{q}_i} F_i(q) - G(q). \qquad (5.39)$$

Proof. By eqs. (5.37), (5.38) one has

$$\varepsilon \left(\sum_i \frac{\partial L}{\partial q_i} F_i(q) + \sum_i \frac{\partial L}{\partial \dot{q}_i} \frac{dF_i(q)}{dt} - \frac{dG}{dt} \right) = 0.$$

By using the Lagrange equations (2.13) one may cast the above equation in the following form, with Q defined by eq. (5.39),

$$\frac{d}{dt} \left(\sum_i \frac{\partial L}{\partial \dot{q}_i} F_i(q) - G \right) = \frac{dQ}{dt} = 0. \tag{5.40}$$

Example 5.11. Consider a system of N particles interacting via a potential V; by Noether theorem, the invariance properties of the Lagrangian yield the following constants of motion.

1) If the potential V depends only on the relative positions

$$V(\mathbf{x}^1, ..., \mathbf{x}^N) = V(x_i^\alpha - x_i^\beta; \alpha, \beta = 1, ...N; i = 1, 2, 3),$$

then, the Lagrangian is invariant under arbitrary *space translations*: $\delta x_i^\alpha = \varepsilon\, a_i, \quad \delta \dot{x}_i^\alpha = 0.$
Hence, by eq. (5.39), one derives the conservation of the *total momentum*

$$P_i = \sum_\alpha \frac{\partial L}{\partial \dot{x}_i^\alpha} = \sum_\alpha p_i^\alpha.$$

2) If furthermore the potential depends only on the modulus of the relative positions, then the Lagrangian is invariant under arbitrary infinitesimal *space rotations*, with axes denoted by the unit vectors $\mathbf{n}$:

$$\delta x_i^\alpha = \varepsilon\, \varepsilon_{ijk} n_j\, x_k^\alpha, \quad \delta \dot{x}_i^\alpha = \varepsilon\, \varepsilon_{ijk} n_j\, \dot{x}_k^\alpha,$$

(the sum over repeated indices j, k being understood).
Hence, by Noether theorem one derives the conservation of the *total angular momentum*:

$$L_i = \sum_\alpha \frac{\partial L}{\partial \dot{x}_i^\alpha} = \sum_\alpha \varepsilon_{ijk}\, x_j^\alpha\, p_k^\alpha.$$

3) If the potential depends on the relative positions, under the infinitesimal *(Galilei) transformations of the Lagrangian variables*:

$$\delta x_i^\alpha = \varepsilon w_i \, t, \quad \delta \dot{x}_i^\alpha = \varepsilon \, w_i,$$

the Lagrangian is invariant up to a total derivative:

$$\delta L = \delta T = \varepsilon \sum_\alpha \frac{d \left(m^\alpha \, x_i^\alpha \, w_i \right)}{dt} + O(\varepsilon^2) = \varepsilon \, \frac{d(M X_i \, w_i)}{dt},$$

where $M \equiv \sum_\alpha m^\alpha$ is the total mass and $X_i \equiv \sum_\alpha m^\alpha x_i^\alpha / M$, $i = 1, 2, 3$ denote the position of the center of mass.

Then, from eq. (5.40) one gets that $\mathbf{P} \, t - M \, \mathbf{X}$ is a constant of motion, say $-M \mathbf{X}_0$. This is the statement that the center of mass moves with constant velocity $\mathbf{P}/M$.

5.4.2 *Noether theorem: Hamiltonian formulation

As argued in the previous Chapters, the Hamiltonian formulation qualifies as a more efficient and physically motivated strategy for discussing the dynamics of a mechanical system. In fact, the characterization of a mechanical system is customarily done by directly specifying the Hamiltonian, without passing through the Lagrangian.

The simplest version of the relation between symmetries and conservation laws in the Hamiltonian formulation, adopted in most textbooks, has already been discussed in Section 5.3, namely that the invariance of the Hamiltonian under a one-parameter continuous group $\mathcal{G}$ of canonical transformations, which do not explicitly depend on time, implies the constancy of the corresponding generator. This connection shall be referred to as the standard relation between symmetries and conservation laws.

However, in this way, two important aspects, covered by Noether theorem in the Lagrangian formulation, as discussed above, are left out.

First, *the symmetry of the dynamics is not equivalent to the invariance of the Hamiltonian* (as discussed in Remark 3.1), due to the possible occurrence of boundary terms corresponding to the invariance of the Lagrangian up to a total derivative.

The second issue is the case of explicit *time dependence of the canonical transformations*.

Time independent transformations

In order to discuss the implications of the first issue, we start by discussing the Hamiltonian version of the invariance of the Lagrangian up to a total derivative, under $\mathcal{G}$ transformations which do not explicitly depend on time.

The (canonical) transcription of the infinitesimal transformation of the Lagrangian variables, eqs. (5.37), to the canonical (Hamiltonian) variables q, p is given by

$$\delta q_i = \varepsilon F_i(q), \quad \delta p_i = -\varepsilon \sum_j p_j \frac{\partial F_j(q)}{\partial q_i}, \tag{5.41}$$

since, by eq. (3.5), to first order in ε,

$$p_i' = \sum_j p_j \frac{\partial q_j}{\partial q_i'} = \sum_j p_j \left(\delta_{ij} - \varepsilon \frac{\partial F_j}{\partial q_i} \right).$$

The generator of such a transformation is $\mathcal{Q}_c \equiv \sum_i p_i F_i$ and, if the transformation does not explicitly depend on time, the Hamiltonian is invariant up to a total derivative; in fact, one has

$$\delta H = \varepsilon \{ H, \mathcal{Q}_c \} = -\varepsilon \frac{d \mathcal{Q}_c}{dt} = -\varepsilon \frac{d G(q)}{dt}, \tag{5.42}$$

where eq. (5.40) has been used for the last equality.

Thus, *a symmetry of the dynamics, eqs. (5.37), (5.38), requires the invariance of the Hamiltonian only up to a total derivative.*

On the other side, the function

$$\tilde{\mathcal{Q}}(q, p) \equiv \mathcal{Q}(q, \dot{q}(q, p)) = \sum_j p_j F_j - G(q),$$

with $\mathcal{Q}(q, \dot{q})$ defined by eq. (5.39), is the generator of the following infinitesimal transformation of the canonical variables, (see eqs. (5.32), (4.12)):

$$\tilde{\delta} q_i = \varepsilon \{ q_i, \tilde{\mathcal{Q}} \} = \varepsilon F_i, \quad \tilde{\delta} p_i = \varepsilon \{ p_i, \tilde{\mathcal{Q}} \} = \varepsilon \left(\sum_j -p_j \frac{\partial F_j}{\partial q_i} + \frac{\partial G}{\partial q_i} \right). \tag{5.43}$$

If, as assumed, the $\mathcal{G}$ transformations do not explicitly depend on time (so that $\partial\tilde{\mathcal{Q}}/\partial t = 0$), the Hamiltonian is invariant under such transformations:

$$\delta H = \varepsilon\{H,\ \tilde{\mathcal{Q}}\} = -\varepsilon\frac{d\tilde{\mathcal{Q}}}{dt} = -\varepsilon\frac{d\mathcal{Q}}{dt} = 0, \qquad \tilde{\mathcal{Q}} = \mathcal{Q}_c - G,$$

the conservation of $\tilde{\mathcal{Q}}$ corresponding to eq. (5.40).

As discussed in Remark 3.1, $G(q)$ may be regarded as the generator of a "gauge" transformation and therefore the constant of motion $\tilde{\mathcal{Q}}$ is the sum of the "canonical" generator, $\mathcal{Q}_c$, and the generator $G(q)$ of a gauge transformation,. The invariance of the Hamiltonian may be obtained by combining the transformation (5.41) with a gauge transformation. Thus, in a rather rudimentary form, we encounter the need of adding a gauge term to the canonical generator in order to get conservation, a phenomenon that in modern theoretical physics goes under the name of *anomaly*. The check of the invariance of the Hamiltonian up to a total derivative, eq. (5.42), requires the construction of $\mathcal{Q}_c$ and the question arises of a more direct check.

To this purpose, we first note that, since the dynamics is described by the time evolution of the coordinates $q_i(t)$, (as discussed in Chapter 3, Remark 3.1), a *non-trivial symmetry of the dynamics* must acts non trivially on the $q_i(t)$'s, i.e. $\delta q_i \neq 0$.

Thus, a one-parameter group of non-trivial symmetries of the dynamics is described by canonical transformations $q, p \to q', p'$, with $q' \neq q$, under which the new Hamiltonian function H' differs from the original function H by a total derivative

$$H'(q', p') = H(q', p') - dG'/dt. \tag{5.44}$$

Now, under transformations which do not-explicitly depend on time, the Hamiltonian transforms covariantly, as it does the Lagrangian, i.e. $H'(q', p', t) = H(q(q', p'), p(q', p'), t)$ and the above eq. (5.44) is equivalent to

$$H(q', p', t) - H(q, p, t) = \frac{dG(q)}{dt}, \tag{5.45}$$

the strict analogs of eqs. (5.35), (5.36). This provides a direct characterization of the *time independent canonical transformations which correspond to symmetries of the dynamics* and for infinitesimal transformations one has $\delta H = \varepsilon\, dG/dt$, the analog of eq. (5.38).

Time dependent transformations

For transformations which explicitly depend on time, the Hamiltonian does not transform covariantly (in contrast with the covariance of the Lagrangian), but rather according to eq. (5.15)

$$H'(q',p',t) = H(q,p,t) + \frac{\partial F}{\partial t},$$

so that eq. (5.44) is equivalent to

$$H(q',p',t) - H(q,p,t) - \frac{\partial F}{\partial t} = \frac{dG(q)}{dt}. \tag{5.46}$$

In order to derive the conservation laws following from eq. (5.46), it is convenient to use its infinitesimal form.

Proposition 5.4.2 *Under an infinitesimal time-dependent change of the canonical variables, $q, p \to q', p'$,*

$$\delta q_i = \varepsilon \frac{\partial F(q,p,t)}{\partial p_i}, \quad \delta p_i = -\varepsilon \frac{\partial F(q,p,t)}{\partial q_i}, \tag{5.47}$$

one has for $q' = q + \delta q$, $p' = p + \delta p$

$$\Delta H \equiv H'(q',p',t) - H(q(q',p',t),p(q',p',t),t) = \varepsilon \frac{\partial F}{\partial t}. \tag{5.48}$$

Proof. In fact, since $\partial q_i/\partial t = 0 = \partial p_i/\partial t$, eq. (5.13) gives

$$\frac{\partial \delta q_i}{\partial t} = \frac{\partial(q_i' - q_i)}{\partial t} = \frac{\partial(H' - H)}{\partial p_i'} = \frac{\partial(H' - H)}{\partial p_i} + O(\varepsilon^2).$$

On the other side, eq. (5.47) gives

$$\frac{\partial \delta q_i}{\partial t} = \varepsilon \frac{\partial\{q_i, F\}}{\partial t} = \varepsilon\{q_i, \frac{\partial F}{\partial t}\},$$

so that $\{q_i, \Delta H - \varepsilon \partial F/\partial t\} = 0$.

Similarly, one gets $\{p_i, \Delta H - \varepsilon \partial F/\partial t\} = 0$. Hence, $\Delta H - \varepsilon \, \partial F/\partial t$ has vanishing Poisson brackets with the whole set of canonical variables and therefore must be a constant (which does not contribute to the dynamics).

As a consequence of the above Proposition, a one-parameter group of symmetries of the dynamics implies that the Hamiltonian is invariant under the corresponding infinitesimal canonical transformation, eqs. (5.47), up to a total derivative according to eqs. (5.46), (5.48), i.e.

$$\delta H \equiv H(q', p', t) - H(q, p, t) = \varepsilon \frac{\partial F}{\partial t} + \varepsilon \frac{d\,G(q)}{dt}. \tag{5.49}$$

Theorem 5.4.3 Noether theorem. Hamiltonian form.
(F.Strocchi, arXiv : 1711.10390v1[physics.class − ph])

To each one-parameter group $\mathcal{G}$ of (non-trivial) symmetries of the dynamics, so that in the Hamiltonian formulation the corresponding infinitesimal transformations of the canonical variables, eq. (5.47), with $\delta q_i \neq 0$, yield the invariance of the Hamiltonian up to a total derivative, eq. (5.49), there corresponds a conserved quantity

$$\mathcal{Q} \equiv F + G.$$

Proof. One one side, one has

$$\delta H = \varepsilon \left(\frac{\partial H}{\partial q_i} \frac{\partial F}{\partial p_i} - \frac{\partial H}{\partial p_i} \frac{\partial F}{\partial q_i} \right) = \varepsilon \{H, F\} = -\varepsilon \left(\frac{dF}{dt} - \frac{\partial F}{\partial t} \right),$$

and, on the other side, by eq. (5.49), one has

$$\delta H = \varepsilon \frac{\partial F}{\partial t} + \varepsilon \frac{d\,G(q)}{dt}.$$

Hence, it follows that

$$-\frac{dF}{dt} = \frac{dG(q)}{dt}.$$

i.e. $\mathcal{Q} = F + G$ is a constant of motion.
The non triviality of $\mathcal{Q}$ follows from the assumption $\delta q_i = \varepsilon \{q_i, F\} \neq 0$, whereas $\{q_i, G\} = 0$.

The above Theorem generalizes the result discussed after eq. (5.43) for transformations which do not explicitly depend on time.

Quite generally, the constant of motion corresponding to the invariance of the Hamiltonian up to a total derivative dG/dt, under infinitesimal transformations of the canonical variables generated by F, eqs. (5.47), is the sum of F and the generator G of the "gauge" transformation (see Remark 3.1)

$$q_i \to q_i, \quad p_i \to p_i - \partial G / \partial q_i.$$

This is clearly displayed by Example 3.6, considering for simplicity the case with $V = 0$. The transformations

$$q' = p - wt =, \; p' = p,$$

are generated by $F = -pt$, and by applying eq. (5.47) one gets

$$\Delta H = H(q', p') - H(q(q', p', t), p(q', p', t)) + \varepsilon w \, p = \varepsilon w \, p = \varepsilon \, w \, \frac{d(mq)}{dt}.$$

Hence, the Hamiltonian is invariant up to the total derivative $d(mq)/dt$ and the constant of motion is

$$\mathcal{Q} = -pt + m \, q.$$

Clearly for the case of N particles in $\mathbf{R}^3$, with an interaction potential which depends only on the relative positions, one obtains the same constants of motion derived in Example 5.11, using the Lagrangian formulation. In fact, the canonical generators $\mathbf{F} = -\mathbf{P} \, t$ of the infinitesimal canonical transfromations are not conserved and constants of motion are obtained by adding the generators $M \, \mathbf{X}$ of gauge transformations.

It is worthwhile to remark that $\mathcal{Q}_i = -P_i \, t + M \, X_i$ generates the following Hamiltonian Galilei transformations of the canonical variables

$$\delta x_i^\alpha = -w_i \, t, \quad \delta p_i^\alpha = -w_i.$$

6
Small oscillations

6.1 Equilibrium configurations. Stability

In the following, we consider conservative systems interacting with a continuous (regular) potential.

A configuration point $Q_0 = (q_1^0, ..., q_n^0)$ is an *equilibrium point* if the forces vanish at this point, i.e. if $\partial V/\partial q_i|_{Q=Q_0} = 0$ (and the initial velocities vanish).

Q_0 is a *stable equilibrium point* if configuration points initially close to Q_0 remain close to Q_0 under time evolution, if the initial velocities are sufficiently small.

More precisely, if for any given $\delta > 0$, there is a correspondent $\varepsilon > 0$, such that for all initial configurations $Q(0), \dot{Q}(0)$, with $|Q(0) - Q_0| < \varepsilon$, $|\dot{Q}(0)| < \varepsilon$, one has $|Q(t) - Q_0| < \delta$, $|\dot{Q}(t)| < \delta$, $\forall t$.

This is very simply illustrated by a point mass in one space dimensions subject to a positive quadratic potential $V(x) = \frac{1}{2}\omega^2(x - x_0)^2$. Clearly, if the initial energy $E = \frac{1}{2}m\dot{x}^2 + V(x)$ is smaller than ε, by energy conservation, since both the kinetic energy and the potential are positive definite, one has for all times

$$\dot{x}(t)^2 < 2E/m < 2\varepsilon/m, \quad (x(t) - x_0)^2 < 2E/\omega^2 < 2\varepsilon/\omega^2.$$

More generally, the equilibrium points are stationary points of the potential and the stability is decided by the positivity of the second derivative of the potential in that equilibrium point.

The original version of this chapter was revised: Display equation has been corrected.
An erratum to this chapter can be found at https://doi.org/10.1007/978-3-319-73761-4_8

© Springer International Publishing AG 2018
F. Strocchi, *A Primer of Analytical Mechanics*, UNITEXT for Physics,
https://doi.org/10.1007/978-3-319-73761-4_6

Actually, as stated by **Dirichlet theorem** *the local (isolated) minima of the potential are stable equilibrium points.*

In fact, if Q_0 is a local isolated minimum of the potential, there is a sufficiently small neighborhood $\mathcal{N}$ of Q_0, such that $V(Q) \geq V(Q_0)$, $\forall Q \in \mathcal{N}$. Moreover, by the continuity of V, for an initial position $Q(0)$ sufficiently close to Q_0, one may obtain $0 < V(Q(0)) - V(Q_0) < \varepsilon/2$. On the other hand, at any later time t, by the conservation of the energy, one has

$$T(\dot{Q}(t)) + V(Q(t)) - V(Q_0) = T(\dot{Q}(0)) + V(Q(0)) - V(Q_0).$$

Hence, if the initial velocities are taken sufficiently small, one may have $T(\dot{Q}(0)) < \varepsilon/2$, so that $T(\dot{Q}(t)) + [V(Q(t)) - V(Q_0)] < \varepsilon$. Since both terms are positive, this implies $T(Q(t)) < \varepsilon$, and therefore $V(Q(t)) - V(Q_0) < \varepsilon$. This indicates that $Q(t)$ remains close to Q_0.

For a more precise argument, consider the open ball B_ε of radius ε in the configuration space of the Lagrangian coordinates $Q, \dot{Q}$, centered in the point $Q_0, \dot{Q} = 0$.

Since the boundary ∂B_ε is a compact set, by Weierstrass theorem, the (continuous positive) energy function $E(Q, \dot{Q})$ defined there, has a (positive) absolute minimum λ, i.e. $E(Q, \dot{Q}) \geq \lambda$, $\forall Q, \dot{Q} \in \partial B_\varepsilon$.

Then, if $B_\delta \subset B_\varepsilon$ denotes the ball on which $E(Q, \dot{Q}) < \lambda$, the initial data in B_δ cannot give rise to trajectories which get out of B_ε, since by energy conservation $E(Q(t), \dot{Q}(t)) = E(Q(0), \dot{Q}(0)) < \lambda$ and therefore the trajectory $Q(t), \dot{Q}(t)$ cannot cross ∂B_ε, on all the points of which $E \geq \lambda$.

Example 6.1 Consider the system described in Example 1.2 and further discussed at the beginning of Section 2.4.

Choosing now $z = 0$ at the bottom of the circle and as Lagrangian coordinate the angle ϕ, so that $z = r(1 - \cos\phi)$, one has

$$L = \tfrac{1}{2} mr^2 \dot{\phi}^2 + \tfrac{1}{2} mr^2 \omega^2 \sin^2\phi - mgr(1 - \cos\phi) = T' + V_{eff},$$

$$T' \equiv \tfrac{1}{2} mr^2 \dot{\phi}^2, \quad V_{eff} \equiv -\tfrac{1}{2} mr^2 \omega^2 \sin^2\phi + mgr(1 - \cos\phi).$$

Thus, the equilibrium points are the solutions of

$$0 = dV_{eff}/d\phi = mr\sin\phi(-\omega^2 r\cos\phi + g),$$

i.e. $\phi = 0, \pi$, and $\phi = \arccos(g/\omega^2 r)$, which has solutions only if $g/\omega^2 r \leq 1$.

The stability is decided by

$$d^2 V_{eff}/d\phi^2 = m\,r[\cos\phi(-\omega^2 r\cos\phi) + \omega^2 r\sin^2\phi].$$

The point $\phi = \arccos(g/\omega^2 r)$ is stable, since at this point $d^2 V_{eff}/d\phi^2 = \omega^2 \sin^2\phi > 0$. If $g/\omega^2 r > 1$, the point $\phi = 0$ is a stable point since then $d^2 V_{eff}/d\phi^2|_{\phi=0} = m\,r(-\omega^2 r + g) > 0$. The point $\phi = \pi$ is always unstable since $d^2 V_{eff}/d\phi^2|_{\phi=\pi} = -m\,r(\omega^2 r + g) < 0$.

* **Example 6.2.** *The spherical pendulum.*

   In spherical coordinates (see eq. (2.27)) the Lagrangian for the spherical pendulum reads

$$\mathcal{L} = \tfrac{1}{2}m\,l^2(\dot\theta^2 + \sin^2\theta\,\dot\varphi^2) + mg\,l\cos\theta. \tag{6.1}$$

The angular momentum $p_\varphi = m\,l^2\sin^2\theta\,\dot\varphi$ is conserved. If at the initial time $p_\varphi = 0$, then the motion takes place in a plane $\varphi = constant$ and one recovers the case of the plane pendulum. We then consider the case $\mu \equiv p_\varphi/(m\,l^2) \neq 0$. Then the effective Lagrangian becomes

$$\mathcal{L}_{eff} = \tfrac{1}{2}m\,l^2\,\dot\theta^2 - V_{eff}, \quad V_{eff} \equiv -m\,l(g\cos\theta + \tfrac{1}{2}l\mu^2\sin^{-2}\theta).$$

The equilibrium points are the solutions of

$$dU_{eff}/d\theta = (g/l)\sin\theta - \mu^2\cos\theta/\sin^3\theta = 0, \quad U_{eff} \equiv V_{eff}/ml^2.$$

This equation reduces to a quartic equation for $x \equiv \sin^2\theta$ of the form $ax^4 + x - 1 = 0$ and an acceptable solution $(0 < x \leq 1)$ exists, corresponding to $\theta = \theta_m$, with $0 < \theta_m < \pi/2$. (In the simplest case $a = 1$, the solution is $x \simeq 0.5654$, $\theta_m = 0.8510$). Such an equilibrium point is stable because

$$\frac{d^2 U_{eff}}{d\theta^2}\Big|_{\theta=\theta_m} = \frac{g}{l}\cos\theta_m + \frac{\mu^2}{\sin^2\theta_m}\left(1 + 3\frac{\cos\theta_m}{\sin^2\theta_m}\right) > 0,$$

since $\cos\theta_m = (g/l)\mu^{-2}\sin^4\theta_m > 0$.
For initial data with energy $E > V_{eff}(\theta_m)$ the angle $\theta(t)$ may oscillate between the two values θ_1, θ_2 corresponding to the solutions of the equation $E - V_{eff}(\theta) = 0$, corresponding to the points where $\dot\theta(t) = 0$.

6.2 Small oscillations

As discussed above, the motion in the neighborhood of a stable equilibrium point is approximately governed by the second derivatives of the potential. Thus, we are led to consider the motion described by quadratic Lagrangians:

$$L = \tfrac{1}{2} \sum_{i=1}^{N} M_{ij}\, \dot{q}_i\, \dot{q}_j - V_{ij}\, q_i\, q_j, \tag{6.2}$$

where M_{ij} is a positive definite symmetric matrix and V_{ij} is a symmetric matrix.

There are two possible approaches to the analysis of the motion corresponding to the above Lagrangian.

The **first method** consists in transforming the above Lagrangian to a diagonal form, i.e. involving only terms of the form $\dot{q}_i^2$ and q_i^2.

To this purpose, we may make a linear orthogonal transformation $\mathcal{T}$ of the Lagrangian variables $Q = (q_1, ...q_N) \to \mathcal{T}Q$, $\dot{Q} \to \mathcal{T}\dot{Q}$, such that, in terms of the new variables $\dot{Q}^{\mathcal{T}} \equiv \mathcal{T}\dot{Q}$, the kinetic energy takes a diagonal form

$$T = \tfrac{1}{2} \sum_{i=1}^{N} \tau_i (\dot{Q}_i^{\mathcal{T}})^2.$$

Such a transformation exists because M is a real symmetric matrix and the τ_i are positive numbers because M is strictly positive (i.e. $\operatorname{Tr} M > 0$, $\det M > 0$).

The next step is to rescale the new variables $Q^{\mathcal{T}} \to \widehat{Q^{\mathcal{T}}} \equiv SQ^{\mathcal{T}}$, with $(SQ^{\mathcal{T}})_i \equiv \sqrt{\tau_i}Q_i^{\mathcal{T}} \equiv \widehat{Q^{\mathcal{T}}}_i$, so that

$$T = \tfrac{1}{2} \sum_{i} (\dot{\widehat{Q^{\mathcal{T}}}}_i)^2.$$

Finally, we make an orthogonal transformation $\widehat{Q^{\mathcal{T}}} \to \tilde{Q} \equiv V\widehat{Q^{\mathcal{T}}}$ which brings V to diagonal form

$$V = \tfrac{1}{2} \sum_{i} k_i (\tilde{Q}_i)^2,$$

and obviously does not affect the form of the kinetic energy.

In conclusion, in terms of the new variables the Lagrangian describes N uncoupled systems

$$L = \tfrac{1}{2}\sum_i [(\dot{\tilde{Q}}_i)^2 + k_i \tilde{Q}_i^2]. \tag{6.3}$$

Clearly, the time evolution of $\tilde{Q}$ is immediately given by the eigenvalues k_i of the matrix defined by $V(\widehat{Q^{\mathcal{T}}})$ and the corresponding eigenvectors allow to determine the time evolution of the original Q_i's as linear combinations of the $\tilde{Q}_i(t)$'s

$$Q(t) = \mathcal{T}^{-1}S^{-1}\mathcal{V}^{-1}\tilde{Q}(t). \tag{6.4}$$

If $k_i > 0$, $\tilde{Q}_i(t)$ describes an harmonic oscillator, with frequency $\omega_i \equiv \sqrt{k_i}$, and it is called a **normal mode**.

If $k_i = 0$, $\tilde{Q}_i(t)$ describes a free evolution

If $k_i < 0$, one has $\tilde{Q}_i = Ae^{-\omega_i t} + Be^{+\omega_i t}$, with $\omega_i = \pm i\sqrt{|k_i|}$, and the constants A, B are determined by the initial conditions.

Example 6.3 Consider a triatomic molecule, e.g. H_2S (hydrogen-sulfide). The equilibrium configuration is a linear one with the sulfur atom in the central point and the two hydrogen atoms at the two ends. Discuss the corresponding small oscillations.

It is a good approximation to neglect transverse motion (i.e. to assume that the configuration of the three atoms keeps being linear) and to describe the motion around the equilibrium point in the quadratic harmonic approximation of the potential, neglecting a direct interaction between the two hydrogen atoms:

$$V(x_1, x_2, x_3) = \tfrac{1}{2}k[(x_1 - x_3)^2 + (x_2 - x_3)^2], \tag{6.5}$$

where x_1, x_2, x_3 denote the displacements from the equilibrium positions of the two hydrogen atoms and of the sulfur atom.

The kinetic energy matrix is already in diagonal form (so that according to the first method discussed above $\mathcal{T} = 1$) and it is reduced to the identity by the following rescaling $x_3 \to \alpha_{-1}x_3 \equiv X$, $\alpha \equiv \sqrt{m/M}$, where M denotes the mass of the sulfur atom and m the mass of the hydrogen atom.

Then, the matrix defined by $V(x_1, x_2, X)$ has the following form (the overall factor $k/2$ is omitted)

$$\hat{V} = \begin{pmatrix} 1 & 0 & -\alpha \\ 0 & 1 & -\alpha \\ -\alpha & -\alpha & 2\alpha \end{pmatrix}$$

Its eigenvalues are the solutions of

$$0 = \det(\hat{V} - \lambda \mathbf{1}) = (1-\lambda)\lambda(\lambda - 2\alpha^2 - 1),$$

i.e. $\lambda_0 = 0$, $\lambda_1 = 1$, $\lambda = 1 + 2\alpha^2$.

The corresponding eigenvectors are, respectively, (omitting normalization constants)

$$v_0 = \begin{pmatrix} 1 \\ 1 \\ \alpha^{-1} \end{pmatrix}; \quad v_1 = \begin{pmatrix} 1 \\ -1 \\ 0 \end{pmatrix}; \quad v_{1+2\alpha^2} = \begin{pmatrix} 1 \\ 1 \\ -2\alpha \end{pmatrix} \equiv v_3.$$

Clearly, the time evolution of v_0 is a free evolution (no harmonic force acting); it corresponds to a variable $\tilde{Q}_0$ which is the following combination of the original variables:

$$\tilde{Q}_0 = x_1 + x_2 + \alpha^{-1}X = x_1 + x_2 + (M/m)x_3,$$

i.e. (apart to an overall normalization constant) to the position x_G of the center of mass. In fact, it is easy to directly check by using the Lagrange equations that $\ddot{x}_G = 0$.

The time evolution of v_1 is an harmonic oscillation with frequency $\tilde{\omega}_1 = 1$; it corresponds to the linear combination $\tilde{Q}_1 = x_1 - x_2$, which oscillates with frequency $\omega_1 = \sqrt{k/m}$. This means that the two hydrogen atoms move with 180 degrees out of phase, while the sulfur atom remains fixed.

Finally, the harmonic oscillation of v_3 describes the time evolution of the variable $\tilde{Q}_3 = x_1 + x_2 - 2x_3$. Such a linear combination of the original variables corresponds to the two hydrogen atoms moving in phase and the sulfur atom moving in the opposite direction; the oscillation frequency is $\sqrt{k/m}(1 + 2m/M)$.

An alternative **second method** is to directly find the oscillation frequencies by using the Lagrange equations (sum over repeated indices is understood). From the Lagrangian (6.2) one has:

$$M_{ij}\ddot{q}_j = V_{ij}q_j, \tag{6.6}$$

and then one looks for solutions of the form $\tilde{q}_i^\alpha = \operatorname{Re} A_i^\alpha e^{i\omega_\alpha t}$.
By the linearity of the equations, one may start with $A_i^\alpha e^{i\omega^\alpha t}$ and take the real part at the end. The existence of non-trivial solutions of this form requires that the corresponding frequencies ω_α are such that

$$\det(-\omega_\alpha^2 M + V) = 0. \tag{6.7}$$

Once the frequencies ω_α are found the corresponding modes are obtained as solutions of the following eigenvalue equation (sum over j)

$$(-\omega_\alpha^2 M_{ij} + V_{ij})\, A_j^\alpha = 0. \tag{6.8}$$

Example 6.3. We adopt the second method for solving the problem of Example 6.3.

The mass matrix $\mathcal{M}$ and the potential matrix $\mathcal{V}$ are

$$\mathcal{M} = \begin{pmatrix} m & 0 & 0 \\ 0 & m & 0 \\ 0 & 0 & M \end{pmatrix}; \quad \mathcal{V} = \begin{pmatrix} k & 0 & -k \\ 0 & k & -k \\ -k & -k & -2k \end{pmatrix};$$

and correspondingly eq. (6.7) gives

$$(-\omega_\alpha^2 m + k)\,\omega_\alpha^2\,(\omega_\alpha^2 m\, M - k\, M - 2\,k\,m) = 0.$$

The frequencies ω_α which solve such an equation are:

$$\omega_0^2 = 0, \quad \omega_1^2 = k/m, \quad \omega_2^2 = (k/m)(1 + 2m/M).$$

For each ω_α^2, $\alpha = 0, 1, 2$, is easy to solve the eigenvalue equation (6.8), and in this way, as expected, one recovers all the results obtained by the previous method.

Example 6.4. (*Double pendulum*) Consider a pendulum with mass m_2 attached to another pendulum with mass m_1, both attached to rigid massless wires of length l and discuss the plane motion in the approximation of small oscillations.

Denoting by φ_1, φ_2 the angles describing the deviations of the two wires from the vertical line, the velocities v_1, v_2 of the two masses are given by $v_1 = l\,\dot{\varphi}_1$, $v_2 = l\,(\dot{\varphi}_2 + \dot{\varphi}_1)$. The (gravitational) potential V of the two masses is given by

$$V(\varphi_1, \varphi_2) = l\,m_1 g \cos\varphi_1 + l\,m_2 g(\cos\varphi_1 + \cos\varphi_2) \simeq$$

$$\simeq \tfrac{1}{2}l\,g[(m_1 + m_2)\varphi_1^2 + m_2\varphi_2^2],$$

the last equality corresponding to the approximation of small oscillations.

Then the mass and potential matrices are

$$\mathcal{M} = \tfrac{1}{2}l^2 \begin{pmatrix} m_1 + m_2 & m_2 \\ m_2 & m_2 \end{pmatrix}; \quad V = \tfrac{1}{2}l\,g \begin{pmatrix} m_1 + m_2 & 0 \\ 0 & m_2 \end{pmatrix}.$$

By applying the (first) method of successive diagonalizations, it is convenient to first reduce the matrix V to a multiple of the identity by the following transformationof the Lagrangian variables

$$q_1 = \sqrt{m_1 + m_2}\,l\,\varphi_1, \quad q_2 = \sqrt{m_2}\,l\varphi_2,$$

which leads to the following form of the Lagrangian:

$$L = \tfrac{1}{2}(\dot{q}_1^2 + \dot{q}_2^2) + \sqrt{(m_2/m_1 + m_2)}\dot{q}_1\,\dot{q}_2 - \tfrac{1}{2}(g/l)(q_1^2 + q_2^2).$$

Next, one may diagonalize the mass matrix, by the following transformation

$$\sqrt{2}Q_1 = q_1 + q_2, \quad \sqrt{2}Q_2 = q_1 - q_2,$$

leading to

$$L = \tfrac{1}{2}(\mu_1\,\dot{Q}_1^2 + \mu_2\,\dot{Q}_2^2) - \tfrac{1}{2}(g/l)(Q_1^2 + Q_2^2), \quad \mu_{1,2} = 1 \pm \sqrt{m_2/(m_1 + m_2)}.$$

Thus, the frequencies of oscillations are

$$\omega_1^2 = (g/l)/\mu_1, \quad \omega_2^2 = (g/l)/\mu_2.$$

By the alternative (second) method, one applies eq. (6.7), which in this case gives the following equation for the frequencies w_α, $\alpha = 1, 2$,

$$w_\alpha^4 \, m_2 = (w_\alpha^2 + (g/l))^2 (m_1 + m_2).$$

The solutions are

$$w_\pm^2 = \pm \frac{g}{l} \sqrt{\frac{m_1 + m_2}{m_2}} \left(1 \mp \sqrt{\frac{m_1 + m_2}{m_2}} \right)^{-1}$$

which coincide with $w_{1,2}^2$ given above.

7

*The common Poisson algebra of classical and quantum mechanics

7.1 Dirac Poisson algebra

As discussed in Chapters 4, 5, the algebra of canonical variables with the (Lie) product defined by the Poisson bracket provides the general structure for the formulation of Hamiltonian classical mechanics and may be considered as its backbone.

Actually, most of the general issues, like time evolution, transformations of canonical variables, symmetries and constants of motion etc. have a simple and neat formulation in terms of such an algebraic structure.

Clearly, Dirac must have had in mind the power and effectiveness of the classical canonical structure in formulating the quantization rules in such a way to reproduce as closely as possible the general algebraic properties of Hamiltonian mechanics (*canonical quantization*).

In fact, in this way, important physical quantities, like e.g. the Hamiltonian, the momentum and the angular momentum keep being the generators of, respectively, time translations, space translations and space rotations, provided that their action is given by commutators rather then by the Poisson brackets.

© Springer International Publishing AG 2018
F. Strocchi, *A Primer of Analytical Mechanics*, UNITEXT for Physics,
https://doi.org/10.1007/978-3-319-73761-4_7

Amazingly, as it may *a posteriori* appear, at a *formal level* the quantum revolution may be reduced and fully accounted for, merely by the replacement of the classical Poisson brackets $\{\,,\,\}$ by commutators (**Dirac canonical quantization**)

$$[\hat{q}_i,\,\hat{q}_j] = 0 = [\hat{p}_i,\,\hat{p}_j], \quad [\hat{q}_i,\,\hat{p}_j] = i\hbar\{q_i,\,p_j\} = i\hbar\delta_{ij}, \qquad (7.1)$$

where $[\,,\,]$ denotes the commutator and $\hat{q}$, $\hat{p}$ the quantum version of the classical canonical variables q, p.

The issue of further understanding and justifying such a strong relation between classical and quantum mechanics was of great concern for Dirac, as discussed in Chapter IV of his book *The Principles of Quantum Mechanics*.

Dirac suggested to explain the above relation between classical and quantum mechanics by abstracting the following algebraic structure as common to both classical and quantum mechanics.

To this purpose, Dirac starts by considering the (real regular) functions $f(q, p)$ of the canonical variables and the corresponding real associative algebra $\mathcal{A}$, with identity **1**, (see Section 4.2), i.e. the real vector space with a (not necessarily commutative) product

$$f, g \to f\,g, \quad (f\,g)(q, p) \equiv f(q, p)\,g(q, p), \quad \mathbf{1}\,f = f\mathbf{1} = f, \qquad (7.2)$$

which is clearly associative: $f\,(g\,h) = (f\,g)\,h$. Hereafter, such a product will be called the *basic product*.

Dirac further equips such an algebra with a *Lie product*

$$f, g \to \{f,\,g\}_L \equiv \{f,\,g\}_{PB}, \qquad (7.3)$$

where $\{.\,,\,.\}_{PB}$ denotes the classical Poisson brackets.

In this way $\mathcal{A}$ becomes a Poisson algebra, briefly called the **Dirac Poisson algebra** and as abstract algebra is denoted by $\mathcal{A}_D$. .

If, one considers the *abstract algebra* $\mathcal{A}_D$ as the common Poisson algebra of classical and quantum mechanics, respectively defined by the classical canonical variables and by the corresponding quantum canonical variables, the only distinctive feature is that in the classical realization/representation of such an abstract algebra the basic product (7.2) is commutative, whereas it is not commutative in the quantum case.

A priori, it would seem that the commutativity of the basic product is largely free and one may ask which additional ingredients have to added in order to fix the commutator.

The very important property discovered by Dirac is that in a Poisson algebra $\mathcal{A}$ the following algebraic identity (called **Dirac identity**) holds, which establishes a link between the Lie product $\{.,.\}$ and the commutator $[A, B] \equiv AB - BA$ (defined in terms of the basic product), for any $A, B \in \mathcal{A}$.

Proposition 7.1.1 *If $\mathcal{A}$ is Poisson algebra the following properties hold:*
*1) (**Dirac identity**)*

$$[A, C]\{B, D\} = \{A, C\}[B, D], \quad \forall A, B, C, D \in \mathcal{A}; \tag{7.4}$$

2) the commutator and the Lie product commute

$$[A, B]\{A, B\} = \{A, B\}[A, B]; \tag{7.5}$$

3) if there exists a pair C, D, such that $\{C, D\}$ is invertible, as assumed in the following, then,

$$[[C, D], \{C, D\}^{-1}] = 0,$$

$$[[C, D]\{C, D\}^{-1}, \{A, B\}] = 0, \quad \forall A, B \in \mathcal{A}; \tag{7.6}$$

4) if also $\{A, B\}$ is invertible, then

$$[A, B]\{A, B\}^{-1} = [C, D]\{C, D\}^{-1} = \{C, D\}^{-1}[C, D] \equiv Z, \tag{7.7}$$

5) Z relates the commutator to the Lie product, in the sense that $\forall E, F, G, H \in \mathcal{A}$

$$[E, F] = Z\{E, F\}, \quad [Z, \{G, H\}] = 0 = [Z, [G, H]]. \tag{7.8}$$

6) Z commutes with all the elements of $\mathcal{A}$, i.e. it is a central variable:

$$[Z, A] = 0, \quad \forall A \in \mathcal{A}. \tag{7.9}$$

Proof.

1) The proof cleverly exploits the Leibniz rule satisfied by the Lie product, with respect to the basic product. In fact, by applying it first to the product AB and then to the product CD, one has, respectively,

$$\{AB, CD\} = \{A, CD\}B + A\{B, CD\} =$$

$$= \{A, C\}DB + C\{A, D\}B + A\{B, C\}D + AC\{B, D\}; \qquad (7.10)$$

$$\{AB, CD\} = \{AB, C\} + C\{AB, D\} =$$

$$\{A, C\}BD + A\{B, C\}D + C\{A, D\}B + CA\{B, D\}. \qquad (7.11)$$

Then, by subtracting eq. (7.10) from eq. (7.11) one gets eq. (7.4)

$$\{A, C\}[B, D] + [C, A]\{B, D\} = 0.$$

2) Eq. (7.4) trivially implies that, for any pair $A, B \in \mathcal{A}$, their commutator and the Lie product commute

$$[A, B]\{A, B\} = \{A, B\}[A, B].$$

3) Since $\{C, D\}$ is invertible, eq. (7.4) applied to C, D, C, D, gives

$$[C, D] = \{C, D\}[C, D]\{C, D\}^{-1} = [C, D] + \{C, D\}[[C, D], \{C, D\}^{-1}].$$

Hence, $\{C, D\}[[C, D], \{C, D\}^{-1}] = 0$ and, since $\{C, D\}$ is invertible, $[[C, D], \{C, D\}^{-1}] = 0$.

Furthermore, from eq. (7.4) applied to A, B, C, D and to C, D, A, B, using that $\{C, D\}$ is invertible, one respectively gets

$$[A, B] = \{A, B\}[C, D]\{C, D\}^{-1},$$

$$\{C, D\}^{-1}[C, D]\{A, B\} = [A, B].$$

Then, subtracting one equation from the other and using that $[C, D]$ and $\{C, D\}^{-1}$ commute, one obtains eq. (7.6).

4) Equation (7.7) follows from eq. (7.4) applied to A, B, C, D, using that both $\{A, B\}^{-1}$ and $\{C, D\}^{-1}$ commute with the corresponding commutators.

5) The first of eqs. (7.8) follows from eq. (7.4) applied to E, F, C, D, using that $\{C, D\}$ is invertible and eq. (7.6).

Furthermore, Z commutes with the basic product, since, by eq. (7.6) it commutes with the Lie product, to which the basic product is related by eq. (7.8):

$$[Z, [E, F]] = [Z, Z\{E, F\}] = Z[Z, \{E, F\}] = 0.$$

6) Quite generally, by applying eq. (7.4) to A, B, EC, D, and using the Leibniz rule, one has

$$[A, B]\{EC, D\} = \{A, B\}[EC, D] = \{A, B\}(E[C, D] + [E, D]C),$$

On the other hand, by the Leibniz rule

$$[A, B]\{EC, D\} = [A, B](E\{C, D\} + \{E, D\}C),$$

so that, by comparing the right hand sides of the two equations and using eq. (7.4), one has

$$[A, B]E\{C, D\} = \{A, B\}E[C, D], \qquad (7.12)$$

and, by using eqs. (7.8)

$$\{A, B\}ZE\{C, D\} = \{A, B\}EZ\{C, D\}.$$

Choosing $A = C, B = D$, by the existence of $\{C, D\}^{-1}$, one gets $[Z, E] = 0, \forall E \in \mathcal{A}$.

Equation (7.5) was proved by Dirac, (in his book *The Principles of Quantum Mechanics*, Chapter IV), who also provided (semi-heuristic) arguments for the validity of eqs. (7.8), (7.9), without specifying the (mathematical) conditions for their validity, clearly pointed out in the above Proposition.

Actually, eqs. (7.8, (7.9) do not hold for general Poisson algebras. In particular, the existence of pairs C, D such that $\{C, D\}$ is invertible fails if the Poisson algebra is generated by C^∞ functions of compact support.

Equation (7.9) was proved by Farkas and Letzter for prime Poisson algebras (D. R. Farkas and G. Letzter, J. Pure Appl. Algebra, **125**, 155-190,(1998)).

The next step, in Dirac attempt to explain the relation between classical and quantum mechanics, on the basis of a common (underlying) algebraic structure, was to consider irreducible representations of the Dirac Poisson algebra, so that the central variable Z gets represented by a multiple of the identity.

Then, the deep structural relation between classical and quantum mechanics is that they both correspond to irreducible representations of the *same* Dirac Poisson algebra, being distinguished by the representative values taken by the central variable Z, respectively $Z = 0$ and $Z = i\hbar$.

More precisely, according to Dirac strategy, denoting by f, g the representative of elements of $\mathcal{A}_D$ in the classical representations, actually given by functions $f(q, p)$, $g(q, p)$ of the classical canonical variables, and by u_f, u_g the corresponding elements in the quantum mechanical representation, one obtains the following algebraic relation between classical an quantum canonical variables

$$u_{f+g} = u_f + u_g, \quad u_1 = 1, \tag{7.13}$$

$$[u_f, u_g] = i\hbar\, u_{\{f,g\}_{PB}}. \tag{7.14}$$

Unfortunately, eq. (7.14) is incompatible with the so-called Von Neumann rule $u_{g(f)} = g(u_f)$; it is even incompatible with linearity, if irreducibility of the resulting algebra, or related conditions, are assumed (for the detailed discussion of such inconsistencies see S.T. Ali and M. Englis, Rev. Math. Phys. **17**, 391 (2005), Section 1).

In conclusion, Dirac argument for explaining the deep relation between classical and quantum mechanics at the level of the canonical structure is mathematically inconsistent and the Lie algebraic structure of the quantum variables cannot be defined by eq. (7.14).

A modified Dirac strategy is discussed in the next Section, following G. Morchio and F. Strocchi, Lett. Math. Phys. **86**, 135-150 (2008); Rep. Math. Phys. **64**, 33-48 (2009).

7.2 A common Poisson algebra of classical and quantum mechanics

The above inconsistencies indicate that the Dirac Poisson algebra encodes too much structure of the classical case, not shared by the algebra of quantum canonical variables. Therefore, for the identification of an algebra common to classical and quantum mechanics, one must start with the minimal requirements expected to be shared by both.

To this purpose, as a first step, one may argue that in both cases one may introduce an (abstract) algebra which describes the "positions" of the system. More precisely, one may consider the polynomial associative algebra $\mathcal{A}_q$ generated by the C^∞ real functions of the "coordinates" q, including the identity function $\mathbf{1}$.

The vector space relations are the (obvious) standard ones and the (basic) associative product is defined as before (see eq. (7.2)): $\forall f, g \in \mathcal{A}_q$

$$f, g \to f\,g, \quad (f\,g)(q) \equiv f(q)\,g(q), \quad \mathbf{1}\,f = f\mathbf{1} = f.$$

For simplicity, in the following we shall consider $\mathbf{R}^N$ as the position space (even if one may consider a general C^∞ manifold $\mathcal{M}$).

The next step it to realize that on such an algebra one may define infinitesimal translations, and that their action define real vector fields and a corresponding Lie structure:

$$\delta_i f = \partial f / \partial q_i = \{f,\, p_i\}, \quad \forall f \in \mathcal{A}_q \tag{7.15}$$

where p_i denotes the vector field in the i-th direction.

These premises allow to introduce the abstract *free polynomial (real) associative algebra* $\mathcal{A}$ *generated by the coordinates* q_i *and by the* p_i's.

The Lie product is extended to $\mathcal{A}$, *exclusively* by assuming the Leibniz rule, the linearity in both arguments of the product and antisymmetry, never using commutativity of the (basic) associative product.

Since, the Lie product is meant to be related to translations, its extension to $\mathcal{A}$ is done by posing

$$\{q_i,\, q_j\} = 0. \tag{7.16}$$

In this way, one obtains a **Poisson algebra** (still denoted by) $\mathcal{A}$, which is different from the Dirac Poisson algebra $\mathcal{A}_D$. Our proposal is that the so defined $\mathcal{A}$ may be taken as a *common Poisson algebra of classical and quantum mechanics.*

Since $\{q_i, p_i\} = \mathbf{1}$, $\forall i$, the Lie product of all such pairs is invertible, so that the conditions of Proposition 7.1.1 are satisfied and, as a consequence, the commutator is related to the Lie product by a central variable (no summation over repeated indices is understood):

$$Z \equiv [q_i,\, p_i]\{q_i,\, p_i\}^{-1} = [q_i,\, p_i] = [q_j,\, p_j]. \tag{7.17}$$

Hence,

$$[q_i,\, q_j] = 0, \quad [q_i,\, p_j] = Z\,\{q_i,\, p_j\} = Z\,\delta_{ij}. \tag{7.18}$$

In the irreducible representations of $\mathcal{A}$, more generally in the representations in which the central variables are represented by multiples of the identity (the so called factorial representations), $Z = \lambda\mathbf{1}$, λ a complex number.

Then, $Z = 0$ gives the commutative classical case, whereas $Z \neq 0$ corresponds to the non-commutative quantum case; these are the only kinds of factorial representations of $\mathcal{A}$ (the scale of Z being clearly undetermined).

The intrinsic geometric meaning of the algebraic structure of $\mathcal{A}$, namely the existence of translations in the space of coordinates, replaces the somewhat *ad hoc* assumptions by Dirac, eqs. (7.13), (7.14), on the basis of a claimed classical analogy. No such a classical analogy may be invoked for relating classical and quantum mechanics; the only relation is that they correspond to *inequivalent* realizations of the Poisson algebra $\mathcal{A}$ defined above, which consists of free polynomials of the coordinates q_i and the generators of translations p_i.

The inequivalence of the two realizations precludes the existence of a mapping between them (as for inequivalent representations of a Lie algebra) and explains the obstructions for Dirac strategy.

By construction, the Poisson algebra $\mathcal{A}$ is generated by elements which are real, i.e. they satisfy a reality or hermiticity condition

$$q_i^* = q_i, \quad p_i^* = p_i.$$

The $*$ operation has a unique extension to $\mathcal{A}$ satisfying $(A B)^* = B^* A^*$ and $\{A, B\}^* = \{A^*, B^*\}$, $\forall A, B \in \mathcal{A}$. Thus,

$$[A, B]^* = -[A^*, B^*], \quad Z^* = -Z. \tag{7.19}$$

$Z \neq 0$, equivalently the non-commutativity of the basic product requires a *complex structure,,* i.e. $Z = i \hbar \mathbf{1}$, $\hbar$ a real number. This explains on general grounds the origin and need of a complex structure in the formulation of quantum mechanics, in contrast with the real realization of classical mechanics.

Similar results may be obtained in the general case of a configuration space corresponding to a C^∞ manifold $\mathcal{M}$.

In this case, the basic ingredient is the existence of local translations defined by vector fields of compact support in $\mathcal{M}$, acting on $C_0^\infty(\mathcal{M})$, the C^∞ functions of compact support in $\mathcal{M},$. Such an action defines a Lie structure and a Lie product. The common Poisson algebra is generated by $C_0^\infty(\mathcal{M})$, by the identity function and by the polynomials of vector fields (for details see G. Morchio and F. Strocchi, refs. above).

Erratum to: Small oscillations

Erratum to:
Chapter 6 in: *A Primer of Analytical Mechanics Prime*
UNITEXT for Physics, https://doi.org/10.1007/978-3-319-73761-4

In the original version of the book, the first unnumbered display equation, after Eq. 6.1, in Example 6.2 has to be corrected in Chapter 6. The erratum chapter and the book have been updated with the change.

The updated online version for this chapter can be found at
https://doi.org/10.1007/978-3-319-73761-4_6

F. Strocchi, *A Primer of Analytical Mechanics*, UNITEXT for Physics,
https://doi.org/10.1007/978-3-319-73761-4_8

A
Appendix. Problems with solutions

One of the main lesson from the following problems is to convince the reader that the Lagrange formulation (with minimal Lagrangian coordinates) greatly simplify the solutions with respect to the Newtonian approach in Cartesian coordinates, with the occurrence of constraint forces.

The reader is strongly invited to explicitly check the advantage of the Lagrangian approach by comparing the two methods of solutions in the various problems below.

1. A box of mass m slides on a frictionless inclined plane of mass M, which forms an angle θ with respect to the horizontal plane. The inclined plane is the upper surface of a body P of mass M which rests on the frictionless horizontal plane. Describe the motion and determine the constraint force on the box.

Solution. Clearly, the problem reduces to a two dimensional problem in the plane defined by the horizontal line, x-axis, and the vertical line z-axis (the y coordinate is not relevant). The body P is assumed to lie on the horizontal plane, with its wedge on the left; its position may be conveniently described by the position X of the wedge, initially put at the origin of the x-axis.
Without loss of generality, the box may be considered as pointlike with coordinates x, z (related to the position of its center of mass) and define $s \equiv x - X$, so that $z = s \tan \theta$; X, s may be taken as (minimal) Lagrangian coordinates, so that one does not need to introduce constraint forces.

© Springer International Publishing AG 2018
F. Strocchi, *A Primer of Analytical Mechanics*, UNITEXT for Physics,
https://doi.org/10.1007/978-3-319-73761-4

Then, the Lagrangian is

$$L = \tfrac{1}{2}M\dot{X}^2 + \tfrac{1}{2}m(\dot{x}^2 + \dot{z}^2) - m\,g\,z =$$

$$= \tfrac{1}{2}M\dot{X}^2 + \tfrac{1}{2}m[(\dot{s} + \dot{X})^2 + \dot{s}^2\tan^2\theta] - m\,g\,s\,\tan\theta.$$

One Lagrange equation is

$$\frac{d}{dt}\frac{\partial L}{\partial \dot{X}} = \frac{d}{dt}\left((M+m)\dot{X} + m\,\dot{s}\right) = \frac{\partial L}{\partial X} = 0,$$

which implies the momentum conservation law, so that if both the box and the plane are initially at rest

$$(M+m)\dot{X} + m\,\dot{s} = 0. \tag{A.1}$$

The second Lagrange equation is

$$\frac{d}{dt}\frac{\partial L}{\partial \dot{s}} = m\,[(1 + \tan^2\theta)\ddot{s} + \ddot{X}] = \frac{\partial L}{\partial s} = -mg\tan\theta.$$

The momentum conservation law allows to express $\ddot{X}$ in terms of $\ddot{s}$; then one obtains

$$\left(\frac{M}{M+m} + \tan^2\theta\right)\ddot{s} = -g\tan\theta.$$

Equivalently, since $x = s + X$, by the momentum conservation $\ddot{x} = M\ddot{s}/(M+m)$,

$$\ddot{x} = -gM\sin\theta\,\cos\theta/(M + m\sin^2\theta).$$

The time evolution of s is therefore a uniformly accelerated motion, with initial conditions $s(0) = s_0$, $\dot{s}(0) = 0$. The time evolution of X is derived by the above conservation law.

The reaction R of the plane on the box is derived from the Newton equation for the box acceleration along the x-axis

$$m\ddot{x} = -R\sin\theta.$$

Hence,

$$R = -m\,M\,g\,\cos\theta/(M + m\sin^2\theta).$$

It is worthwhile to remark that when there are easily available as many constants of motion as the Lagrangian degrees of freedom, one may simply exploit them to solve the dynamical problem; to this purpose one has to by express the constants of motion in terms of the (minimal) Lagrangian coordinates, so that the possible (holonomic) constraints do not enter in the conservation laws.

This is the case of the problem discussed above, where one has both the *momentum conservation*

$$(M + m)\dot{X} + m\,\dot{s} = 0,$$

and the *energy conservation*

$$\tfrac{1}{2}M\dot{X}^2 + \tfrac{1}{2}m[(\dot{s} + \dot{X})^2 + \dot{s}^2 \tan^2 \theta] + m\,g\,s\,\tan\theta = m\,g\,s_0\,\tan\theta.$$

The momentum conservation allows to express $\dot{X}$ in terms of $\dot{s}$ in the equation of the energy conservation

$$\frac{1}{2}\left(\frac{M}{M+m} + \tan^2\right)\dot{s}^2 + g\,s\,\tan\theta = g\,s_0\,\tan\theta.$$

This is the energy of a one-dimensional particle initially at rest under the action of a positive linear potential.

Indeed, the time derivative of the above energy conservation immediately gives the equation of motion derived before for $\ddot{s}$.

2. A homogeneous rope of length l and of total mass m may slide on a table and initially a fraction z_0 of the rope hangs over the edge of the table. Determine the motion of the rope.

Solution. We denote by z the fraction of the rope which hangs over the edge of the table at time t and by $\rho = m/l$ the density of the rope. Then, the kinetic energy has the two contributions corresponding to the fraction of the rope which slides on the table and to the fraction hanging over the edge:

$$T = \tfrac{1}{2}(l - z)\rho\dot{z}^2 + \tfrac{1}{2}z\rho\,\dot{z}^2.$$

The potential is

$$V = \rho g \int_z^0 s\,ds = -\tfrac{1}{2}\rho g\,z^2.$$

Hence, the Lagrangian is

$$L = T - V = \tfrac{1}{2} m \, \dot{z}^2 + \tfrac{1}{2} \rho g z^2,$$

and the Lagrange equation is

$$\frac{d}{dt} \frac{\partial L}{\partial \dot{z}} = m \, \ddot{z} = \frac{\partial L}{\partial z} = z \, \rho \, g.$$

The general solution is

$$z(t) = A e^{t \, g/l} + B e^{-t \, g/l},$$

and the initial conditions $z(0) = z_0$, $\dot{z}(0) = 0$, give

$$z(t) = z_0 (e^{t \, g/l} + e^{-t \, g/l}).$$

Again, since one has only one degree of freedom, the energy conservation immediately provides the dynamical law:

$$0 = \frac{d}{dt}(T + V) = \dot{z}[m \, \ddot{z} - z \, \rho \, g] = 0.$$

3. Determine the equation of motion of a pendulum, whose support undergoes a preassigned horizontal motion. Discuss the conditions under which one may consider the small oscillations of the pendulum.

Consider the case in which the support on mass M may move freely on an horizontal line and determine the frequencies of the small oscillations.

Solution. We denote by $X(t)$ the horizontal preassigned time evolution of the support, by l the length of the (massless) rod to which a point mass m is hanged and by $x(t)$, $z(t)$ the horizontal and vertical coordinates of the mass point.
Denoting by θ the angle formed by the rod with respect to the vertical line, one has

$$x(t) = X(t) + l \, \sin \theta, \quad z(t) = l \, \cos \theta;$$

$$\dot{x} = \dot{X} + l \, \cos \theta \, \dot{\theta}, \quad \dot{y} = -l \, \sin \theta \, \dot{\theta}.$$

The Lagrangian is

$$L = \tfrac{1}{2}m(l^2\dot\theta^2 + \dot X^2 + 2l\dot\theta\dot X\,\cos\theta) - mgl(1 - \cos\theta).$$

The Lagrange equation of motion is

$$\frac{d}{dt}\frac{\partial L}{\partial\dot\theta} = ml\frac{d}{dt}(l\dot\theta + \dot X(t)\cos\theta) = ml(l\ddot\theta + \ddot X\,\cos\theta - \dot\theta\dot X\sin\theta) =$$

$$\frac{\partial L}{\partial\theta} = -ml(\dot\theta\dot X\sin\theta + g\sin\theta),$$

i.e.

$$l\ddot\theta + \ddot X\,\cos\theta + g\sin\theta = 0.$$

Thus, if the support moves with a constant velocity, so that $\ddot X(t) = 0$, one recovers the usual equation of motion of the pendulum.
More generally, if $\ddot X(t) \ll g$, one may try a small angle approximation, leading to

$$\ddot\theta + \ddot X/l + (g/l)\,\theta = 0.$$

This is the equation for a driven oscillator.
A simple case is when the preassigned motion of the support is periodic. The above condition $\ddot X(t) \ll g$ requires $X(0)\omega^2/l \ll g/l \equiv \omega_0^2$, where ω is the frequency of the periodic motion of the support and $\omega_0 = \sqrt{g/l}$ is the frequency of the pendulum, in the familiar case of fixed support. In this problem the energy is not conserved and one cannot exploit the corresponding conservation equation to directly derive the equation of motion, as in the previous problems.

If the support is free, one must add to the Lagrangian the kinetic energy of the support, with motion $X(t)$ to be determined. Then, one gets the same Lagrange equation for θ

$$l\ddot\theta + \ddot X\,\cos\theta + g\sin\theta = 0,$$

and the following equation for $X(t)$

$$\frac{d}{dt}[(M + m)\,\dot X + ml\dot\theta\,\cos\theta] = 0,$$

i.e.

$$\frac{d}{dt}[X(t) + \frac{m}{M + m}l\,\sin\theta] = constant.$$

Then, if initially both the support and the pendulum are at rest, and $X(0) = 0$ one has

$$X(t) = -\frac{m\,l}{M+m}\sin\theta.$$

For small oscillations, $\sin\theta \sim \theta$ and $\ddot{X} = -l\,m/(M+m)\ddot{\theta}$. Thus,

$$\ddot{\theta} + \omega^2\theta = 0, \quad \omega^2 \equiv \frac{M+m}{m}\frac{g}{l}.$$

Thus, both masses oscillate with the same frequency ω, moving in opposite directions.

A limiting case is when $\theta(0) = 0 = \dot{\theta}(0)$, $X(0) = 0$, $\dot{X}(0) \neq 0$; then both the support and the point mass m move with the same velocity, in the same direction.

4. A particle of mass m is initially at rest on the top of a vertical circle and later starts sliding on the circle. Find the equation of motiuon and determine when it starts flying off the circle.

Solution. We denote by θ the angle which the position of the particle on the circle forms with horizontal x-axis, so that the Cartesian coordinates of the particles are $x = R\cos\theta$, $z = R\sin\theta$. Then, the Lagrangian is

$$L = \tfrac{1}{2}mR^2\dot{\theta}^2 - m\,g\,R\sin\theta$$

and the Lagrange equations are

$$R\ddot{\theta} = -g\cos\theta.$$

Putting $\theta = \pi/2 - \phi$, one has

$$\ddot{\varphi} = (g/R)\sin\varphi$$

and the initial motion for $\varphi(0) = 0$, $\dot{\varphi}(0) \ll 2\sqrt{g/R}$, as long as $\varphi(t)$ keeps being small, is given by

$$\varphi(t) = \tfrac{1}{2}\sqrt{\frac{R}{g}}\,\dot{\varphi}(0)[e^{\sqrt{g/R}\,t} - e^{-\sqrt{g/R}\,t}].$$

The particle will start flying off when the reaction $\mathcal{R}$ of the circle will vanish.

From the equation for the acceleration normal to the circle $a_n = -R^2 \dot{\theta}^2$,

$$ma_n = \mathcal{R} - mgR\sin\theta,$$

$\mathcal{R} = 0$ when $R\dot{\theta}^2 + g\sin\theta = 0$.
Since the energy conservation (for the initial data $\theta(0) = 0$, $\dot{\theta}(0) = 0$) reads

$$R\dot{\theta}^2 = 2g - g\sin\theta,$$

the above equation $R\dot{\theta}^2 + g\sin\theta = 0$ becomes $2 - 3\sin\theta = 0$, which yields $\sin\theta = 2/3$, $\theta = 41.8°$.

5. (Atwood machine) Two masses, m_1 and m_2 are connected by an inextensible massless string of length l over an ideal massless pulley. Determine the motion and the tension of the string.

Consider the case in which the pulley has radius R and moment of inertia I.

Solution. Putting the origin of the vertical z-axis such that the (vertical) position of the pulley is l and denoting by z_1, z_2 the positions of the two masses, one has $z_1 + z_2 = l$ and the Lagrangian is

$$\tfrac{1}{2}(m_1 + m_2)\dot{z}_1^2 - g(m_1 - m_2)z_1 - g\,m_2\,l.$$

Hence, the Lagrange equation is

$$\frac{d}{dt}\frac{\partial L}{\partial \dot{z}_1} = (m_1 + m_2)\ddot{z}_1 = \frac{\partial L}{\partial z_1} = (m_2 - m_1)z_1$$

and the time evolution corresponds to a uniformly accelerated motion. The tension T may be derived from the Newton equation for z_1:

$$m_1\ddot{z}_1 = -g\,m_1 + T, \quad T = 2\,g m_1 m_2/(m_1 + m_2).$$

Taking into account the moment of inertia of the pulley amounts to an additional term in the kinetic energy, corresponding to the rotational energy of the pulley, namely $\tfrac{1}{2}I\,\omega^2$, $\omega = \dot{z}_1/R$.
Hence, the Lagrange equation becomes

$$(m + 1 + m_2 + I/R^2)\ddot{z}_1 = (m_2 - m_1)z_1.$$

6. A homogeneous disc of mass m and radius R is hanging form a (horizontal) plane through a string which is wrapped around it and it is allowed to fall. Determine the motion and the string tension, under the assumption that the string always keeps being vertical (a sort of Yo-Yo motion).

Consider the case in which the support undergoes a preassigned vertical motion.

Solution. The configuration of the disc is described by the (vertical) position z of its center of mass and we denote by θ the rotation angle of the disc, so that $\theta = z/R$. Therefore, the kinetic energy is

$$\tfrac{1}{2}m\dot{z}^2 + \tfrac{1}{2}I\dot{\theta}^2, \quad I = \tfrac{1}{2}mR^2,$$

and the Lagrangian is

$$L = (3/4)m\dot{z}^2 + mgz.$$

Hence, the Lagrange equation is

$$(3/2)m\ddot{z} = m\,g,$$

describing a uniformly acceleration motion. The tension T may be determined by the Newton equation for the center of mass

$$m\ddot{z} = m\,g + T.$$

It is instructive to solve the problem by using the Newton equations for z and θ as well as by exploiting the energy conservation.
If the support undergoes a preassigned vertical motion $Z(t)$, then the vertical position of the disc is now $z' = z + Z$, with $z = R\theta$, as above. Then, the Lagrangian is

$$\tfrac{1}{2}m(\dot{z} + \dot{Z})^2 + (1/4)m\dot{z}^2 + m\,g\,(z + Z),$$

and the Lagrange equation is

$$\ddot{z} = \frac{2}{3}(g - \dot{Z}).$$

7. A sphere of radius r and mass m subject to gravity is rolling without slipping inside a hollow fixed cylinder of radius R. Assume that the motion takes place in the plane orthogonal to the axis of the cylinder and passing through the center of mass of the sphere.

Discuss the motion and determine the period for small oscillations. Consider the case in which the sphere may also roll in the direction of the axis of the cylinder.

Solution. We choose Cartesian coordinates x, y on plane of motion, with origin on the cylinder axis and we denote by θ the angle such that the position of the center of mass of the sphere is given by the Cartesian coordinates $x = (R - r) \sin \theta$, $y = (R - r) \cos \theta$. The rotation angle of the sphere is denoted by φ and its moment of inertia by $I = 2m\,r^2/5$.

Then, the Lagrangian is

$$L = \tfrac{1}{2}I\dot{\varphi}^2 + \tfrac{1}{2}m\left((R - r)\,\dot{\theta}\right)^2 - m\,g(R - (R - r)\cos\theta) =$$
$$= (7/10)m\,(R - r)^2\,\dot{\theta}^2 - m\,g(R - (R - r)\cos\theta).$$

The Lagrange equation is

$$(7/5)m\,(R - r)^2\ddot{\theta} = -m\,g(R - r)\sin\theta,$$

i.e.

$$\ddot{\theta} = -\frac{5g}{7(R - r)}\sin\theta.$$

Hence, the motion is similar to that of a pendulum and the frequency ω of small oscillations is

$$\omega = \left(\frac{5g}{7(R - r)}\right)^{1/2}.$$

If the sphere may also move in the z-direction, one has two additional terms in the kinetic energy corresponding to the velocity in the z-direction of the center of mass, $\tfrac{1}{2}m\,\dot{z}^2$, and to the rotational energy $\tfrac{1}{2}Iw^2$ associated to the angular velocity w orthogonal to the cylinder axis, so that $\dot{z} = r\,w$.

Since the potential is independent of z, one has

$$\frac{d}{dt}\frac{\partial L}{\partial \dot{z}} = (7/5)\,m\,\ddot{z} = 0,$$

i.e. the motion in the z-direction is a free motion.

Index

angular momentum
  coupling with angular velocity, 13
anomaly, 78

Bloch equations, 54

canonical structure, 48
canonical transformation, 37
  Poisson brackets, 52
canonical transformations, 36
  continuous group, 67
  generating functions, 57
  generating function, 58
  generator, 67
canonical variables, 30
  cylindrical, 49
  infinitesimal variations, 69
  spherical, 49
complex structure, 101
conjugate momenta, 20
constant of motion, 43
constraint forces, 1, 4
coordinate transformation, 32, 36
covariance of the Lagrangian
  coordinate transformations, 9
cyclic variables, 23

degrees of freedom, 5
Dirac identity, 95
Dirac Poisson algebra, 94
Dirichlet theorem, 84

Eastern deviation of a falling object, 14
electromagnetic interaction, 16

electron in a central potential, 20
energy conservation, 27
equilibrium configurations, 83
equilibrium point, 83
extended canonical transformations, 65

fictitious forces, 3, 4, 12
Foucault pendulum, 15

Galilei transformation
  conservation laws, 81
  Hamiltonian variables, 63
Galilei transformations, 73
  Hamiltonian variables, 81
  Lagrangian variables, 76
gauge transformation, 40, 78
generalized force, 8
generalized potential, 16

Hamilton equations, 30
Hamiltonian
  change of coordinates, 32
  harmonic oscillator, 28
  invariance up to total derivative, 78
  particle in a central potential, 28
  atomic electron in a magnetic field, 34
  change for a rotating frame, 34
  invariance, 60
  invariance up to total derivative, 41
  particle in an e.m. potential, 28
Hamiltonian and energy, 27, 28
Hamiltonian formulation, 27
Hamiltonian function, 27
Hamiltonian Galilei transformation, 64

© Springer International Publishing AG 2018
F. Strocchi, *A Primer of Analytical Mechanics*, UNITEXT for Physics,
https://doi.org/10.1007/978-3-319-73761-4

holonomic constraints, 10

invariance of Lagrangian, 23

Jacobi identity, 47
Jacobi theorem, 47

Kronecker symbol δ_{ij}, 48

Lagrange equations, 5, 7–9
  covariance, form invariance, 9
Lagrangian, 7
Lagrangian coordinates, 5
Lagrangian form of Lorentz equations, 16
Lagrangian form of Newton equations, 7
Lagrangian Galilei transformations, 64
Larmor theorem, 18
Levi-Civita tensor ϵ_{ijk}, 16
Lie product, 47, 94

Noether theorem, 74
  Hamiltonian form, 80
  Hamiltonian formulation, 76
  Lagrangian formulation, 74
non- inertial frame, 12
non-inertial frame, 12
non-inertial frames, 3

particle in a central potential, 53
point mass on a rotating circle, 3
Poisson algebra, 47, 93, 94
  common to classical and quantum
    mechanics, 100

Poisson bracket, 44
Poisson brackets, 43
  canonical transformations, 51
  general structure, 46

quantum mechanics, 94

rotating frame, 12
rotations, 67, 72

scale transformation, 66
scale transformations, 68
simple pendulum, 2
small oscillations, 83, 86
space translations, 72
spherical pendulum, 85
stability, 83
stable equilibrium point, 83
symmetries and conservation laws, 23
symmetries and constants of motion, 74
symmetries of dynamics
  transformation of Hamiltonian, 77
symmetries of the dynamics
  conservation laws, 74

time reversal, 66
time translations, 73
time-dependent canonical transformations,
  60
total derivative, 6
  boundary term, 24
translations, 67
triatomic molecule, 87

Printed in the United States
By Bookmasters